T0252076

PRODUCTION OF COLLOIDAL BIOGENIC SELENIUM AND REMOVAL BY DIFFERENT COAGULATION-FLOCCULATION APPROACHES

Thesis Committee

Thesis Promotor

Prof. Dr. Ir. Piet N.L. Lens
Professor of Environmental Biotechnology
UNESCO-IHE Institute for Water Education
Delft, The Netherlands

Thesis Co-Promotors

Dr. Hab. Eric D. van Hullebusch, Dr. Hab., PhD, MSc
Hab. Associate Professor in Biogeochemistry
University Paris-Est
Paris, France

Dr. Hab. Giovanni Esposito, Dr. Hab., PhD, MSc
Hab. Associate Professor
University of Cassino and the Southern Lazio
Cassino, Italy

Other Members

Dr. Erkan Şahinkaya
Istanbul Medeniyet University
Istanbul, Turkey

Prof. Stefan Uhlenbrook
UNESCO-IHE Institute for Water Education
Delft, The Netherlands

Dr. Hab. Paul Mason
Utrecht University
Utrecht, The Netherlands

This research was conducted under the auspices of the Erasmus Mundus Joint Doctorate Environmental Technologies for Contaminated Solids, Soils, and Sediments (ETeCoS^3) and The Netherlands Research School for the Socio-Economic and Natural Sciences of the Environment (SENSE).

Joint PhD degree in Environmental Technology

Docteur de l'Université Paris-Est
Spécialité : Science et Technique de l'Environnement

Dottore di Ricerca in Tecnologie Ambientali

Degree of Doctor in Environmental Technology

Thèse – Tesi di Dottorato – PhD thesis

Lucian Staicu

Production of colloidal biogenic elemental selenium and removal by different coagulation-flocculation approaches

Defended on December 19th, 2014

In front of the PhD committee

Prof. Erkan Sahinkaya	Reviewer
Hab. Dr. Mason Paul	Reviewer
Prof. Piet Lens	Promotor
Hab. Dr. Eric van Hullebusch	Co-promoter
Prof. Giovanni Esposito	Examiner
Prof. Stefan Uhlenbrook	Examiner

CRC Press
Taylor & Francis Group
Boca Raton London New York

CRC Press is an imprint of the
Taylor & Francis Group, an **informa** business
A BALKEMA BOOK

European Commission
ERASMUS
MUNDUS

Erasmus Joint doctorate programme in Environmental Technology for Contaminated Solids, Soils and Sediments (ETeCoS3)

First issued in hardback 2018

CRC Press/Balkema is an imprint of the Taylor & Francis Group, an informa business

© 2015, Lucian C. Staicu

All rights reserved. No part of this publication or the information contained herein may be reproduced, stored in a retrieval system, or transmitted in any form or by any means, electronic, mechanical, by photocopying, recording or otherwise, without written prior permission from the publishers.

Although all care is taken to ensure the integrity and quality of this publication and information herein, no responsibility is assumed by the publishers or the author for any damage to property or persons as a result of the operation or use of this publication and or the information contained herein.

Published by:
CRC Press/Balkema
PO Box 11320, 2301 EH Leiden, The Netherlands
e-mail: Pub.NL@taylorandfrancis.com
www.crcpress.com – www.taylorandfrancis.com

ISBN 13: 978-1-138-37332-7 (hbk)
ISBN 13: 978-1-138-02819-7 (pbk)

Table of Contents

List of Tables 9

List of Figures 10

List of Abbreviations 12

Abstract 17

Résumé 18

Samenvatting 19

Sommario 20

Chapter 1. Introduction 24

    1.1.   Introduction   24

    1.2.   Problem description   24

    1.3.   Objectives   25

    1.4.   Structure of the thesis   25

Chapter 2. Treatment technologies for selenium removal 30

    2.1. Introduction   30

    2.2. Natural and anthropogenic sources of selenium   31

    2.3. Selenium contaminated water and wastewater   33

    2.4. Selenium chemistry and toxicity   33

    2.5. Legislation   35

    2.6. Treatment technologies   35

      2.6.1. Physical treatment   35

      2.6.2. Chemical treatment   38

      2.6.3. Biological treatment   44

    2.7. Conclusions and Outlook   50

    2.8. References   51

Chapter 3. *Pseudomonas moraviensis* subsp. stanleyae 60

    3.1. Introduction   60

    3.2. Materials and methods   62

      3.2.1. Media and culture conditions   62

      3.2.2. Isolation of strain #71   62

      3.2.3. Identification of strain #71   63

      3.2.4. Growth test   64

3.2.5. Selenium oxyanions measurement 64

3.2.6. Selenium tolerance 65

3.2.7. Transmission Electron Microscopy (TEM) 65

3.2.8. Inoculation experiment 65

3.2.9. Statistical analysis 66

3.3. Results 66

3.3.1. Phylogenetic analysis 66

3.3.2. Growth 67

3.3.3. SeO_3^{2-} reduction 68

3.3.4. Selenium tolerance 69

3.3.5. Production of red elemental selenium 69

3.3.6. Inoculation 71

3.4. Discussion 72

3.4.1. Phylogenetic analysis 72

3.4.2. Growth 72

3.4.3. SeO_3^{2-} reduction 73

3.4.4. Tolerance to selenite and selenate 73

3.4.5. Production of red elemental selenium 74

3.4.6. Inoculation experiment 75

3.5. Conclusions 76

3.6. References 76

Chapter 4. Electrocoagulation of colloidal biogenic selenium 82

4.1. Introduction 82

4.2. Materials and Methods 84

4.2.1. Reagents and electrodes 84

4.2.2. Biogenic Se(0) production and solution preparation 85

4.2.3. Electrocoagulation set-up 85

4.2.4. Electrocoagulant generation 87

4.2.5. Toxicity Characteristic Leaching Procedure (TCLP) test 88

4.2.6. Analytical methods 88

4.2.7. Calculations 89

4.3. Results and Discussion 89

4.3.1. Characterization of the biogenic Se(0) suspension 89

4.3.2. Electrodissolution of the Al and Fe electrodes 91

4.3.3. Treatment efficiency of electrocoagulation 92

4.3.4. TCLP and supernatant characterization 94

4.3.5. Sediment characterization 96

4.3.6. Electrical energy consumption and process optimization 99

4.4. Conclusions 100

4.5. References 101

Chapter 5. Removal of colloidal biogenic selenium from wastewater 106

5.1. Introduction 106

5.2. Materials and Methods 108

5.2.1. Chemicals and media 108

5.2.2. Production of biogenic red Se(0) 108

5.2.3. Se(0) protein-coating characterization 109

5.2.4. Jar-test experiments 109

5.2.5. Analysis 110

5.2.6. Data analysis 110

5.3. Results 111

5.3.1. Biogenic Se(0) particle characterization 111

5.3.2. Turbidity removal 113

5.3.3. Se(0) surface charge 116

5.3.4. Characterization of sediment 116

5.4. Discussion 116

5.4.1. Turbidity removal 116

5.4.2. Se(0) charge repression 118

5.4.3. Sediment characterization 119

5.4.4. Biogenic Se(0) particle characteristics 120

5.5. Conclusions 121

5.6. References 121

Chapter 6. Conclusions and Perspectives 128

6.1. Environmental impact of selenium 128

6.2. Biogenic Se(0) – Metabolism 129

6.3. Selenium-laden wastewater 130

6.4. Se(0) – Separation 131

6.4.1. Chemical dosing (Coagulation-Flocculation) 132

6.4.2. Electrocoagulation 132

6.5. Recovery and reuse 133

6.5.1. Fertilizers 134

6.5.2. Semiconductors 135

6.5.3. Adsorbent for toxic metals 135

6.6. Perspectives 136

6.7. References 136

Appendix 1: Valorization of PhD research 141

I. Articles 141

II. Conferences 141

III. Summer school presentations 142

IV. Seminars 142

V. Courses and trainings 142

Appendix 2: Curriculum vitae 143

Appendix 3: SENSE Certificate 145

List of Tables

CHAPTER 2

Table 2.1. Concentration of selenium in raw materials and various wastes 34

Table 2.2. ABMet® reactor gradational ORP zones..46

Table 2.3. Selenium removal in full-scale surface flow wetlands 49

CHAPTER 3

Table 3.1. Multi-locus sequence analysis parameters 63

Table 3.2. Growth of *Pseudomonas moraviensis* stanleyae...64

Table 3.3. Fatty Acid Methyl Esters profile of *P. moraviensis* stanleyae...............................66

Table 3.4. Growth of *P. moraviensis* stanleyae and formation of red Se(0) in

LB media under different growth conditions 70

Table 3.5. *Brassica juncea* inoculation with *P. moraviensis* stanleyae...................................71

CHAPTER 4

Table 4.1. Summary of the EC results using Al and Fe sacrificial electrodes 86

Table 4.2. Properties of biogenic Se(0) solution produced by *P. moraviensis* stanleyae 90

Table 4.3. Summary of Se-Fe and Se-Al sediment characteristics 99

CHAPTER 5

Table 5.1. Characteristics of biogenic red Se(0) particles 112

Table 5.2. Properties of biogenic Se(0) solution 113

List of Figures

CHAPTER 1

Figure 1.1. Overview of thesis 26

CHAPTER 2

Figure 2.1. Distribution of Se oxyanions as a function of pH 34

Figure 2.2. Electrocoagulation setup..43

Figure 2.3. Biogenic selenium nanoparticles..45

CHAPTER 3

Figure 3.1. Neighbor-joining tree of *Pseudomonas moraviensis* stanleyae...........................67

Figure 3.2. Aerobic growth of *P. moraviensis* stanleyae 68

Figure 3.3. Aerobic reduction of selenium oxyanions by *P. moraviensis* stanleyae 69

Figure 3.4. Biogenic red Se(0) formation by *P. moraviensis* stanleyae 70

CHAPTER 4

Figure 4.1. Schematic diagram of the electrocoagulation set-up 87

Figure 4.2. Colloidal stability of biogenic red Se(0) produced by
Pseudomonas moraviensis stanleyae 90

Figure 4.3. Variation of theoretical and measured Al and Fe with the electrical charge
passed during electrocoagulation..91

Figure 4.4. Turbidity removal and pH evolution for Fe and Al electrodes 93

Figure 4.5. Leaching behavior of sediment and residuals...95

Figure 4.6. Sediment characterization..98

Figure 4.7. Electrical energy consumption with Al or Fe electrodes.....................................99

Figure 4S. Treatment effectiveness of electrocoagulation 100

CHAPTER 5

Figure 5.1. Biogenic red Se(0) produced by anaerobic granular sludge inoculum
(TEM, Size distribution, SDS-PAGE characterization) 111

Figure 5.2. Variation of the zeta potential of biogenic Se(0) as a function of pH 112

Figure 5.3. Colloidal stability of biogenic Se(0) and removal by filtration and
centrifugation at different speed values 114

Figure 5.4. Removal of colloidal Se(0) (efficiency, pH and zeta potential change)...............115

Figure 5.5. Sediment characterization (volume and CST) 117

CHAPTER 6

Figure 6.1. Biogenic Se(0) production 129

Figure 6.2. Biopolymer-induced colloidal stability of Se(0)................................... 131

Figure 6.3. Separation of colloidal biogenic by electrocoagulation 133

Figure 6.4. Proposed Se recovery and reuse scheme.. 134

List of Abbreviations

ABMet[*]	Advanced Biological Metals Removal
BMM	Basal Mineral Medium
CE	Capillary Electrophoresis
CFU	Colony Forming Unit
CST	Capillary Suction Time
CTP	Comparative Toxicity Potential
DO	Dissolved Oxygen
DTT	Dithiothreitol
EBR	Electro-biochemical reactor
EC	Electrocoagulation
EDX	Energy Dispersive X-ray
EES	Enhanced Evaporation System
EP	Evaporation Pond
EPRI	Electric Power Research Institute
ESEM	Environmental Scanning Electron Microscope
FAME	Fatty Acid Methyl Ester
FBR	Fluidized Bed Reactor
FGD	Flue Gas Desulfurization
Fh	Ferrihydrite
GAC	Granular Activated Carbon
HRT	Hydraulic Retention Time
IEA	International Energy Agency
IC	Ion Chromatography
ICP-OES	Inductively Coupled Plasma-Optical Emission Spectroscopy
KB	King's B growth medium
LB	Lysogeny Broth
LOD	Limit of Detection

MBfR	Membrane Biofilm Reactor
MCE	Microchip Capillary Electrophoresis
MLSA	Multi-locus Sequence Analysis
MS	Mass Spectrometry
NAMC	North American Metals Council
NF	Nanofiltration
NTU	Nephelometric Turbidity Units
OCC	Optimal Coagulant Concentration
OD	Optical Density
ORP	Oxidoreduction Potential
PBS	Phosphate Buffer Saline
PDI	Polydispersity Index
PDMS	Poly(dimethylsiloxane)
PFT	Paint Filter Testing
PTFE	Polytetrafluoroethylene
PZC	Point of Zero Charge
RO	Reverse Osmosis
SDS-PAGE	Sodium Dodecyl Sulfate - Polyacrylamide Gel Electrophoresis
Se(0)	Elemental (zero valent) selenium
SIM	Similarity Index
SRB	Sulfate Reducing Bacteria
TBE	Tris-borate-EDTA buffer
TCLP	Toxicity Characteristic Leaching Procedure
TDS	Total Dissolved Solids
TEM	Transmission Electron Microscope
TOC	Total Organic Carbon
TSS	Total Suspended Solids
UASB	Upflow Aanaerobic Sludge Blanket

USBR	United States Bureau of Reclamation
USEPA	United States Environmental Protection Agency
VSS	Volatile Suspended Solids
XRD	X-ray Diffraction
ZLD	Zero Liquid Discharge
ZVI	Zero Valent Iron

Acknowledgements

I would like to thank Prof. Piet Lens (UNESCO-IHE, the Netherlands), promotor of the thesis, for coordinating and supervising the progress of my research activity and for showing me the way to science.

A special thanks to Dr. Eric van Hullebusch (University of Paris-Est, France), co-promoter of the thesis, for his support and guidance.

I would also like to aknowledge the members of the Environmental Technologies for Contaminated Solids, Soils and Sediments (ETeCoS^3) committee, Dr. Eric van Hullebusch and Dr. Giovanni Esposito (University of Cassino and the Southern Lazio, Italy), for giving me the opportunity to participate to this joint Erasmus Mundus doctorate programme.

I would equally like to acknowledge Prof. Mehmet Oturan (University Paris Est) for his permanent support and collaboration throughout the entire PhD project.

I am highly grateful to my scholarship donor, the European Union through the Erasmus Mundus Joint Doctorate Environmental Technologies for Contaminated Solids, Soils, and Sediments (ETeCoS^3) programme (FPA n°2010-0009).

I would like to thank all the PhD students from ETeCoS^3 programme and its partners.

I would like to acknowledge Fred Kruis and all the members of the Laboratory of UNESCO-IHE and Dr. Nihal Oturan and all the members of Laboratoire Géomatériaux et Environnement (University Paris Est).

I would also like to acknowledge the Fulbright Commission for its financial support during my research stage within Colorado State University (CSU). I am grateful to Prof. Elizabeth Pilon-Smits, Prof. Marinus Pilon, Prof. Christopher Ackerson, Dr. Scott Noblitt and Prof. Charles Henry, all from CSU, for their support and availability. A special thanks to Robin Montenieri and Dr. William Hunter, from United States Department of Agriculture. Prof. Pierre Cornelis from the Free University of Bruxelles and his outstanding student, Lumeng Ye, are equally acknowledged.

Special thanks to my colleagues Rohan Jain, Anna Potysz, Maria Rossaria, Karl Ravet (CSU), Jon Harris (CSU), and Thomas Ni (CSU) for sharing with me many many hours of fertile scientific discussions.

Finally, I would like to express my gratitude to my parents, to whom I owe so much for their unlimited support.

Abstract

Selenium (Se) is a chalcogen element with a narrow window between essentiality and toxicity. The toxicity is mainly related to the chemical speciation that Se undergoes under changing redox conditions. Se oxyanions, namely selenite (Se[IV], SeO_3^{2-}) and selenate (Se[VI], SeO_4^{2-}), are water-soluble, bioavailable and toxic. In contrast, elemental selenium, Se(0), is solid and less toxic. Nevertheless, Se(0) nanoparticles are potentially harmful as particulate Se(0) has been reported to be bioavailable to filter feeding mollusks (e.g. bivalves) and fish. Furthermore, Se(0) is prone to re-oxidation to toxic SeO_3^{2-} and SeO_4^{2-} when discharged into aquatic ecosystems.

Biogenic Se(0) under investigation was produced by the reduction of Na_2SeO_4 under anaerobic conditions using a mixed bacterial inoculum (anaerobic granular sludge) and through the reduction of Na_2SeO_3 under aerobic conditions using a pure microbial culture (*Pseudomonas moraviensis* stanleyae, a novel strain identified and characterized for the first time herein). Both types of Se(0) showed strong colloidal stability within the 2-12 pH range. The colloidal stability is caused by the negatively charged (-15 mV to -30 mV) biopolymer layer covering biogenic Se(0) particles and by their nanometer size. The particle size of Se(0) produced by anaerobic granular sludge ranged between 50 and 300 nm, with an average size of 166 nm. Conversely, the Se(0) particles produced by *Pseudomonas moraviensis* stanleyae are characterized by a lower diameter (~ 100 nm).

The solid-liquid separation potential of Se(0) was assessed by centrifugation, filtration, coagulation-flocculation and electrocoagulation. While all approaches can bring about Se(0) removal from suspension with various degrees of success, electrocoagulation using iron sacrificial electrodes showed the highest removal efficiency (97%). Because biogenic Se(0) is harmful to the environment, appropriate measures must be implemented for the solid-liquid separation using an efficient technology.

Keywords: Selenium; Colloidal; Coagulation-Flocculation; Electrocoagulation; *Pseudomonas moraviensis* stanleyae; Nanoparticles.

Résumé

Le sélénium (Se) est un élément chalcogène avec un domaine de concentration étroit entre essentialité et toxicité. La toxicité est principalement liée à la spéciation chimique du Se qui évolue en fonction des conditions redox du milieu. Les formes oxyanioniques de Se, le sélénite (Se [IV], SeO_3^{2-}) et le séléniate (Se [VI], SeO_4^{2-}), sont solubles dans l'eau, biodisponibles et toxiques. En revanche, le sélénium élémentaire, Se(0), est insoluble et moins toxique. Néanmoins, les nanoparticules du Se(0) sont potentiellement dangereuses pour certains groupes des mollusques (comme les bivalves) et aussi pour les poissons. En outre, lorsque le Se(0) est rejeté dans les écosystèmes aquatiques, sa ré-oxydation jusqu'au sélénite et séléniate peut se produire.

Le sélénium élémentaire d'origine biogénique Se(0) a été produit par la réduction de SeO_4^{2-} dans des conditions anaérobies en utilisant un inoculum microbien mixte (boues granulaires) et par la réduction de SeO_3^{2-} dans des conditions aérobies en utilisant une culture bactérienne pure (une nouvelle souche de *Pseudomonas moraviensis* identifiée et caractérisée pour la première fois dans cette thèse). Les deux types de Se(0) ont montré une forte stabilité colloïdale dans l'écart de pH variant de 2 à 12. La stabilité colloïdale est due à la charge négative (-15 mV à -30 mV) de la couche de biopolymère qui entoure Se(0) et à la taille nanométrique des particules de Se(0). La taille des particules de Se(0) produite par la boue anaérobie granulaire se situait entre 50 et 300 nm, avec une taille moyenne de 166 nm. A l'inverse, les nanoparticules de Se(0) produites par *Pseudomonas moraviensis* stanleyae sont caractérisées par un diamètre plus faible (~ 100 nm).

Compte tenu des risques pour l'environnement engendrés par le relargage du Se(0) biogénique, des mesures appropriées doivent être mises en œuvre pour la séparation solide-liquide en utilisant une technologie efficace. Le potentiel de séparation solide-liquide de Se(0) généré a été évaluée par centrifugation, filtration, coagulation-floculation et électrocoagulation. Alors que toutes les approches présentent des rendements de séparation de Se(0) variables, l'électrocoagulation en utilisant des électrodes sacrificielles de fer a montré l'efficacité d'élimination la plus élevée (97%).

Mots-clés: Sélénium; Colloïde; Coagulation-floculation; Electrocoagulation; *Pseudomonas moraviensis* stanleyae; Nanoparticules.

Samenvatting

Selenium (Se) is een element uit de chalcogene groep met een kleine marge tussen de noodzakelijke en toxische concentraties. De toxiciteit is hoofdzakelijk gerelateerd aan de chemische speciatie die Se ondergaat gedurende variërende redox condities. Se oxyanionen, met name seleniet (Se[IV], SeO_3^{2-}) en selenaat (Se[VI], SeO_4^{2-}), zijn water-oplosbaar, biobeschikbaar en toxisch. In tegenstelling is elementair selenium, Se(0), een vaste stof die minder toxisch is. Se(0) nanodeeltjes zijn mogelijks toch schadelijk, zoals blijkt uit onderzoek dat vast Se(0) biobeschikbaar is voor weekdieren (bvb. bivalven) en vissen. Bovendien staat Se(0) deeltjes bloot aan re-oxidatie tot het meer toxische SeO_3^{2-} of SeO_4^{2-} wanneer ze in aquatische ecosystemen worden geloosd.

Het biogeen Se(0) dat in dit onderzoek werd bestudeerd werd geproduceerd door de reductie van Na_2SeO_4 onder anaërobe condities met behulp van een gemengd bacteriëel inoculum (anaëroob granulair slib) en door de reductie van Na_2SeO_3 onder aërobe omstandigheden door een reincultuur van *Pseudomonas moraviensis* stanleyae (een nieuwe stam voor het eerst geïdentificieerd en gekarakteriseerd in dit werk). Beide types Se(0) laten een sterke colloïdale stabiliteit zien in de pH range 2-12. De colloïdale stabiliteit wordt veroorzaakt door de negative lading (-15 mV tot -30 mV) van de biopolymeer laag die de biogenic Se(0) deeltjes bedekt en door hun nanometer grootte. De deeltjesgrootte van het Se(0) geproduceerd door anaëroob granulair slib variëerde tussen 50 and 300 nm, met een gemiddelde grootte van 166 nm. De Se(0) deeltjes geproduceerd door *Pseudomonas moraviensis* stanleyae hadden een kleinere diameter (~ 100 nm).

De mogelijkheid om de vaste stof - vloeistof van het Se(0) te scheiden via centrifugatie, filtratie, coagulatie-flocculatie en electrocoagulatie is onderzocht. Terwijl alle onderzochte methodes de Se(0) deeltjes uit de suspensie konden verwijderen met verschillende mate van succes, werd de grootste verwijderingsefficiëntie (97%) behaald door electrocoagulatie met behulp van ijzer electrodes. Omdat biogeen Se(0) schadelijk is voor het milieu, moeten gepaste maatregelen genomen worden voor de vaste stof - vloeistof scheiding door het toepassen van een efficiënte scheidingstechnologie.

Sleutelwoorden: Selenium; Colloïdaal; Coagulatie-Flocculatie; Electrocoagulatie; *Pseudomonas moraviensis* stanleyae; Nanodeeltjes.

Sommario

Il Selenio (Se) è un elemento calcogeno a metà tra essenzialità e tossicità. La sua tossicità è principalmente legata alla speciazione chimica che il Selenio subisce al cambiamento delle condizioni di ossido-riduzione. Gli ossianioni del Selenio, rispettivamente selenite (Se[IV], SeO_3^{2-}) e seleniato (Se[VI], SeO_4^{2-}), sono solubili in acqua, biodisponibili e tossici. Di contro, il Selenio elementare Se(0), è solido e meno tossico. Tuttavia, le particelle di Se(0) sono potenzialmente pericolose poiché è stato dimostrato che il loro particolato può essere ingerito dai molluschi (per esempio, bivalvi) e dai pesci. Inoltre, il Se(0) è incline alla ri-ossidazione nelle forme tossiche di SeO_3^{2-} e SeO_4^2, se scaricato negli ecosistemi acquatici.

Il Se(0) biogenico in esame è stato prodotto dalla riduzione del SeO_4^{2-} in condizioni anaerobiche usando una coltura mista come inoculo (fango granulare anaerobico) e attraverso la riduzione del SeO_3^{2-} in condizioni aerobiche mediante l'uso di una coltura pura (un nuovo ceppo di *Pseudomonas moraviensis* identificato e caratterizzato per la prima volta in questo studio). Entrambi i tipi di Se(0) hanno mostrato una forte stabilità colloidale nel range di pH 2-12. La stabilità colloidale è causata dallo strato biopolimerico caricato negativamente (-15 mV to -30 mV) che copre le particelle di Se(0) biogenico e dalla loro dimensione nanometrica. La dimensione della particelle di Se(0) prodotta dal fango granulare anaerobico varia tra 50 e 300 nm, con una dimensione media di 166nm. Al contrario, le particelle di Se(0) prodotte dal *Pseudomonas moraviensis* stanleyae sono caratterizzate da un diametro inferiore (~ 100 nm).

Il potenziale di separazione solido-liquido del Se(0) generato dalla coltura mista è stato misurato tramite centrifugazione, filtrazione, coagulazione-flocculazione e elettrocoagulazione. Sebbene tutti questi metodi siano stati in grado di separare con successo il Se(0) dalla sospensione, l'elettrocoagulazione ha mostrato la maggiore efficienza di rimozione (97%). Poiché il Se(0) biogenico è dannoso per l'ambiente, misure appropriate devono essere implementate per la sua separazione solido-liquido mediante una tecnologia efficiente.

Parole chiave: Selenio; Colloide; Coagulazione-Flocculazione; Elettrocoagulazione; *Pseudomonas moraviensis* stanleyae; Nanoparticelle.

Then you will know the truth,

and the truth will set you free.

(St. John 8:32)

CHAPTER 1

Introduction

Chapter 1. Introduction

1.1.   Introduction

Since its identification in 1817 by the Swedish chemist Jacob Berzelius, more than 150 years passed by until environmental scientists realized the extent of selenium (Se) toxicity to aquatic ecosystems (Eisler, 1985). During the mid-1970s, Lake Belews in North Carolina was affected by Se released by a coal-fired power plant, which resulted in the massive die-off of the local fish populations: 19 out of 20 species were eliminated (Lemly, 2002). In the early 1980s, the Se-laden agricultural drain water discharged in the Kesterson Reservoir, California, severely affected the migratory bird populations and triggered environmental actions (Ohlendorf, 1989). Because of its high bioaccumulative capacity, even low levels of Se present in water can have a devastating effect on the higher levels of the food chain, especially the apex predators (e.g. birds, humans) (Chapman et al., 2010). The apex predators were shown to be exposed to toxic selenium levels 2000 times higher than the water concentration (Wu, 2004). Currently, Se is considered a problematic element due to its narrow window between essentiality and toxicity and also because the trends of energy production based on fossil fuel combustion are increasing.

1.2.   Problem description

Amongst its oxidation states, Se oxyanions, namely selenite (Se[IV], SeO_3^{2-}) and selenate (Se[VI], SeO_4^{2-}), are water-soluble, bioavailable and toxic (Simmons and Wallschlaeger, 2005). Contrary to the old belief that elemental Se, Se(0), is less/non-toxic owing to its solid nature and consequently reduced bioavailability, mounting evidence reveals the deleterious effects induced by Se(0) on filter-feeding mollusks (Luoma et al., 1992; Schlekat et al., 2000) and fish (Li et al., 2008). Furthermore, Se(0) is prone to re-oxidation to toxic SeO_3^{2-} and SeO_4^{2-} when discharged into aquatic ecosystems (Zhang et al., 2004). Biologically produced Se(0) exhibits strong colloidal properties that hampers its solid-liquid separation and extends its persistence in the water column (Buchs et al., 2013, Staicu et al., 2015).

Burning of fossil fuels (e.g. coal and oil) for energy generation and the processing of ores generate Se-rich wastewaters (Lemly, 2004). When these wastewaters are treated through a bioremediation approach (e.g. bioreactors), biogenic Se(0) is produced together with a cleaner wastewater. As the biological treatment of Se-laden wastewaters is expected to expand in the following years, the production of biogenic Se(0) will consequently increase its output.

This thesis aimed to investigate the colloidal properties of biogenic Se(0) and the solid-liquid separation methods that can be employed for the removal of Se(0) as a polishing (post-treatment) step prior to environmental discharge.

1.3.   Objectives

The main objective of this research is to "***investigate the removal of colloidal Se(0) produced by pure and mixed microbial cultures***".

The specific objectives are:

1) To study the production and properties of Se(0) using a microbial mixed culture under anaerobic conditions.
    a. To study the reduction of Se oxyanions to red Se(0)
    b. To study the colloidal properties of biogenic Se(0)
2) The study the solid-liquid separation potential of colloidal Se(0) by coagulation-flocculation.
3) To isolate, identify and characterize an aerobic bacterium with high Se-conversion capacity.
    a) To isolate the bacterial species from a seleniferous environment.
    b) To identify the isolate using a polyphasic approach.
    c) To study the tolerance to SeO_4^{2-} and SeO_3^{2-}.
4) To study the production and properties of Se(0) using a pure culture under aerobic conditions.
    a) To study the conversion rate of SeO_3^{2-} to Se(0).
    b) To study the colloidal properties of Se(0) produced under aerobic conditions.
5) To study the electrochemical treatment of colloidal Se(0) as an alternative to the coagulation-flocculation process.
    a) To compare the performance of two different sacrificial electrode materials (Fe and Al).
    b) To characterize the Fe-Se and Al-Se sludges resulted from the electrochemical treatment.

1.4.   Structure of the thesis

The present dissertation is divided into 6 chapters. The following paragraphs outline the content of the chapters (Figure 1.1).

Chapter 1 presents a general overview of the research, including background, problem description, research objectives and the thesis structure.

Chapter 2 explores the literature about the problem related to Se oxyanions and Se(0) treatment technologies.

Chapter 3 investigates the production of colloidal Se(0) by a novel strain of *Pseudomonas fluorescens*.

Chapter 4 presents the removal of colloidal Se(0) produced by a mixed microbial culture using coagulation-flocculation.

Chapter 5 describes the electrochemical treatment of colloidal Se(0) produced by the *Pseudomonas fluorescens* strain.

Chapter 6 summarizes and draws conclusions on knowledge gained from this study and gives recommendations for future perspective.

All the papers published, accepted or submitted and related to the PhD work are listed in Appendix 1. All the conferences attended during the PhD are also listed.

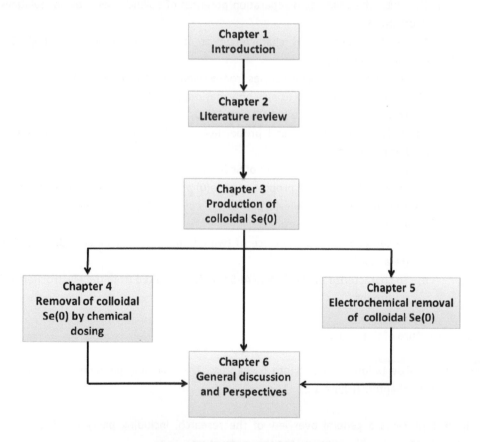

Figure 1.1. Overview of the thesis

1.5.   References

Buchs, B., Evangelou, M.W.H., Winkel, L.H.E., & Lenz, M. (2013). Colloidal properties of nanoparticular biogenic selenium govern environmental fate and bioremediation effectiveness. *Environ. Sci. Technol. 47*(5), 2401-2407.

Chapman, P.M., Adams, W.J., Brooks, M., Delos, C.G., Luoma, S.N., Maher, W.A., Ohlendorf, H.M., Presser, T.S., & Shaw, P. (2010). *Ecological assessment of selenium in the aquatic environments*. SETAC Press, Pensacola, Florida, USA.

Eisler, R. (1985). Selenium hazards to fish, wildlife, and invertebrates: a synoptic review. *US Fish Wild Serv. Biol. Rep.*85.

Lemly, A.D. (2002). Symptoms and implications of selenium toxicity in fish: the Belews Lake case example. *Aquat. Toxicol. 57*(1-2), 39-49.

Lemly, A.D. (2004). Aquatic selenium pollution is a global environmental safety issue. *Ecotox. Environ. Safe. 59*, 44-56.

Li, H., Zhang, J., Wang, T., Luo, W., Zhou, Q., & Jiang, G.(2008). Elemental selenium particles at nano-size (Nano-Se) are more toxic to Medaka (*Oryzias latipes*) as a consequence of hyper-accumulation of selenium: a comparison with sodium selenite. *Aquat. Toxicol. 89*(4), 251-256.

Luoma, S.N., Johns, C., Fisher, N.S., Steinberg, N.A., Oremland, R.S., & Reinfelder, J.R. (1992). Determination of selenium bioavailability to a bivalve from particulate and solute pathways. *Environ. Sci. Technol. 26*, 485-491.

Ohlendorf, H.M. (1989). Bioaccumulation and effects of selenium in wildlife.In Jacobs, L.M. (Ed) *Selenium in agriculture and the environment. Am. S. Agron. S. Sci.* Madison, Wisconsin, series number 23, pp. 133-177.

Schlekat, C.E., Dowdle, P.R., Lee, B.G., Luoma, S.N., & Oremland, R.S. (2000). Bioavailability of particle-associated selenium on the bivalve *Potamocorbila amuresis. Environ. Sci. Technol. 34*, 4504-4510.

Simmons, D.B., & Wallschlaeger, D. (2005). A critical review of the biogeochemistry and ecotoxicology of selenium in lotic and lentic environments. *Environ. Toxicol. Chem. 24*, 1331-1343.

Staicu, L.C., van Hullebusch, E.D., Lens, P.N.L., Pilon-Smits, E.A.H., & Oturan, M.A. (2015). Electrocoagulation of colloidal biogenic selenium. *Env. Sci. Pollut. Res. 22* (4), 3127-3137.

Wu, L. (2004). Review of 15 years of research on ecotoxicology and remediation of land contaminated by agricultural drainage sediment rich in selenium. *Ecotoxicol. Environ. Saf. 57*, 257-269.

Zhang, Y., Zahir, Z.A., & Frankenberger Jr., W.T. (2004). Fate of colloidal-particulate elemental selenium in aquatic systems. *J. Environ. Qual. 33*, 559-564.

CHAPTER 2

Treatment technologies for

selenium removal

This chapter will be submitted for publication as:

Staicu, L.C., van Hullebusch, E.D., & Lens, P.N.L. (2015). Treatment technologies for
selenium-laden wastewaters.

Chapter 2. Treatment technologies for selenium removal

Abstract

Selenium (Se) is an element of concern because it induces toxic effects to biota at relatively low concentrations. A variety of wastewaters mainly related to fossil fuel energy generation and oil refining are in need of effective and economic treatment for selenium removal prior to discharge. The advantages and drawbacks of treating Se-laden wastewaters by physical-chemical and biological technologies are discussed. Some of the main challenges encountered in the treatment of selenium-laden wastewaters are: the presence of Se in several chemical forms, together with the competing anions and cations; the low concentrations of Se, frequently in the order of $\mu g\ L^{-1}$, make it a difficult target for different treatment approaches; the complex wastewater matrix can confound the removal efficiency; various wastewaters (e.g. mining, agriculture) are produced in very large amounts and their source is diffuse; and the treatment generates residual by-products that must be further treated or disposed of safely in order to prevent re-release. The physical-chemical treatment can be effective in bringing the Se levels below the discharge standards but suffers from high investment and operational costs. On the other hand, the biological treatment requires a relatively high footprint and sometimes shows erratic performance. Selenium-charged wastewaters are complex and, therefore, finding a single technology that can treat them efficiently is challenging. Sometimes, a combination of several treatment technologies is required to achieve the discharge limits for selenium and other constituents present in wastewaters.

Keywords: Selenium; Treatment; Wastewater; Bioremediation; Aquatic toxicity.

2.1. Introduction

Selenium (Se) is an element that belongs to the chalcogen group family (together with sulfur, S, and tellurium, Te). Its complex biogeochemistry is due to both abiotic and biotic reactions as well as to its valence states (Fernandez-Martinez & Charlet, 2009; Lenz & Lens, 2009). Selenium is a naturally-occurring element in the Earth's crust but it is present in relatively low abundance (0.05 to 0.5 mg kg^{-1}) (Kabata-Pendias, 2000). Nevertheless, Se varies greatly from region to region and is mainly associated with fossil fuels (e.g. oil, coal, bituminous shale) and phosphate deposits but also found in seleniferous soils (Chapman et al., 2010). When rocks and minerals are mined and processed, or when coal is burned to generate energy, Se is released in the environment. Due to its bioaccumulative capacity, even low levels of selenium build up along trophic levels (i.e. biomagnification) leading to deleterious effects on aquatic ecosystems (e.g. Lake Belews, Kesterson Reservoir) (Lemly, 2002). Furthermore, of particular interest is the very narrow window between essentiality and toxicity elicited by Se (Levander & Burk, 2006).

The first major incident occurred in North Carolina (US) during the mid-1970s, when Lake Belews was affected by Se waste released by a coal-fired power plant resulting in a massive die-off of local fish populations (19 out of 20 species were eliminated) (Lemly, 2002). In the early 1980s, the selenium-laden agricultural drain water discharged into Kesterson Reservoir, San Joaquin Valley (California), severely affected the migratory bird populations and triggered environmental actions (Ohlendorf, 1989).

This chapter summarizes the recent advances in the treatment of selenium-laden wastewaters. Firstly, the physical-chemical treatment methods are revised and discussed. Secondly, biological technologies are presented in detail and compared to the previous section.

2.2. Natural and anthropogenic sources of selenium

Due to close crystallochemical properties, Se can substitute for sulfur in metal-sulfides minerals (e.g. pyrite, chalcopyrite) and in various types of coal. For instance, Chinese stone-coal has been reported to contain up to 6,500 mg kg^{1} Se (Plant et al., 2003). During the Cretaceous, a significant Se enrichment (up to 100 mg kg^{-1}) took place as a result of volcanic gases and dust deposited in marine sediments (Kabata-Pendias, 2000). Consequently, soils that developed on such Cretaceous parent rocks are naturally-rich in selenium (i.e. seleniferous). Notable examples are found in parts of the Central Valley (California, US) and other areas in Western and Central United States.

Selenium is cycled thought different environmental compartments both naturally and by manmade actions (Haygarth, 1994). Natural processes involved in the redistribution of selenium in the environment include terrestrial weathering of rocks and soils, volcanic activity, wildfires and volatilization from soils, vegetation and water bodies (NAMC, 2010). The primary natural sources of selenium in terrestrial systems are clay fractions in sediments, shales, phosphatic rocks, coals and organic deposits (Martinez & Charlet, 2009). Selenium can be released by biogeochemical processes (weathering, rock-water interactions and biological activity) and distributed unevenly over the Earth. Another natural source is the volcanic emissions, selenium being evolved into atmosphere or deposited as fly ash on land and water (Haygarth, 1994). On the other side, ocean is the main contributor to the volatile selenium pool through the algal uptake and biotransformation of soluble selenium oxyanions in seawater (Amouroux et al., 2001). Of no less importance is the transport of selenium associated with wind-eroded particles from land (Haygarth, 1994) or the phytovolatilization of selenium-alkyls by plants (Pilon-Smits & Quinn, 2010).

Although natural processes have their contribution, human activities have greatly accelerated and modified selenium's natural cycle. Combustion of coal for power production releases volatile selenium species and also generates Se enriched residual wastes (fly ash, bottom ash, scrubber ash). The enrichment factor (ratio of initial selenium in coal to selenium in ash) can be 1200 times higher than the parent feed coal (Table 2.1) (Lemly, 2004). About 131 tons of coal combustion ash were generated in 2007 in US alone (Breen, 2009). The ash is sluiced in settling basins characterized by alkaline conditions which promote selenite and selenate formation. This selenium-laden water is periodically discharged into receiving natural waters or can accidently overflow. During one such accident (December, 2008), 1.5 million tons of ash from a Tennessee Valley Authority coal-fired power plant were discharged in the environment as a result of a dike failure (TVA, 2009). Another environmental emerging issue is related to landfill ash disposal which generates toxic landfill leachate and the risk of groundwater contamination (Lemly, 2004). Wen and Carignan (2007) connected the increasing anthropogenic emissions to the onset of the Industrial Revolution (18th century). Moreover, there is a positive correlation between the high coal combustion activity observed in 1940 and 1970 and the selenium accumulation in archived herbage and soil samples from Rothamsted Experimental Station, UK. The data suggest a long-term change in the deposition of selenium, with a gradual increase of about 0.15% yr^{-1} over a period of 100 years. Herbage samples are sensitive to the dynamic of atmospheric composition, reflecting a decrease in selenium concentration consequently to the introduction of Clean Air Act (1956) in the United Kingdom (Haygarth et al., 1994). Considering that currently coal is one of the main energy sources with 40% share of world primary energy, second after oil (IEA, 2014), and taking into consideration the positive trends, an increase in selenium release in the environment is consequently expected.

Phosphate mining and processing are also significant sources of selenium, especially by leaching of surface waste rock dumps (Hamilton, 2004). Selenium being associated with the mineral matrix of sulfide ores (copper, zinc, lead, gold, nickel), is released through the smelting process being readily volatilized and emitted into atmosphere. Here, volatile selenium can adhere to the aerosols and then be deposited on terrestrial or aquatic environments. Depending on the local topography and atmospheric conditions, the aerial plume can travel for long distances, sometimes 100-200 km away from the emission source (Hodson et al., 1984).

Oil industry has its share in selenium release and environmental contamination. Selenium enrichment in crude oils occurred in the same fashion as coal, in fossil organic carbon-rich basins subjected to high temperature and pressure over extended geological eras. Comparing to coal, crude oils contain even higher Se concentrations (Table 2.1) and their refining by- and -end products pose an important risk for the environment. For example, oil processed by refineries surrounding San Francisco Bay severely impacted the local ecosystems (Presser & Luoma, 2006).

Other anthropogenic activities, such as irrigation, increased the amount of selenium in the environment. When irrigation is practiced on soils containing an impervious clay layer ("hardpan"), drains must be constructed to remove water from building up in the top soil. This situation occurred on the seleniferous soils from California and the proposed solution was to divert the drained agricultural runoff to wildlife marshes. As a consequence of the high biaccumulative capacity exhibited by selenium, the aquatic ecosystems were severely affected (Lemly, 1998).

Another source is the use of phosphate fertilizers, as some phosphatic rocks contain high concentrations of selenium (Chapman et al., 2010). A much cited example of selenium application on agriculture soils is Finland. Due to particular characteristics of Finnish soils (young, acidic, weekly weathered and high in adsorptive oxides), selenium shows low bioavailability to crops therefore the need of commercial supplementation emerged (Hartikainen, 2005; Alfthan et al., 2014).

2.3. Selenium contaminated water and wastewater

Wastewaters enriched in Se are mainly generated by mining and oil refining and the coal-fired power generation sector (Table 2.1). Se removal technologies are, however, not so well developed as the energy sector has focused on fly ash handling and flue gas desulfurization systems, whereas mining operations around waste rock and tailing management (NAMC, 2010). Of particular interest is the Se associated with the coal-fired power generation sector. Selenium is a an element that accumulates in coal deposits to levels of 1-1.6 mg Kg^{-1} but can reach concentrations up to 43 mg Kg^{-1} (Yudovich & Ketris, 2005). Moreover, since coal demand is expected to grow at an average rate of 2.3% per year through 2018, the Se release in the environment will continue to increase (IEA, 2013). China, in particular, is increasingly relying on coal to fuel its economic development (IEA, 2013).

2.4. Selenium chemistry and toxicity

Selenium exists in several valence states. The two oxidized forms, or oxyanions, selenite (Se[IV], SeO_3^{2-}) and selenate (Se[VI], SeO_4^{2-}), are water-soluble, bioavailable and toxic (Simmons & Wallschlaeger, 2005). In natural aquatic environments, selenium oxyanions have limited interaction with cations (e.g. Ca^{2+}, Mg^{2+}) and therefore only a small fraction will be removed from solution by precipitation (Seby, 1999) (Figure 2.1). Moreover, unlike metals, they show increased solubility and mobility with increasing pH (Chapman et al., 2010). By contrast, elemental selenium, Se(0), is solid and relatively non-toxic (Schlekat et al., 2000), although concerns exist about the deleterious effects observed for nanoselenium (Zhang E.S. et al., 2005). In addition, Se(0) has been shown to be bioavailable to filter feeders and fish (Luoma et al., 1992; Schlekat et al., 2000; Li et al., 2008) and prone to get reoxidized to SeO_3^{2-} and SeO_4^{2-} when discharged into aquatic environments (Zhang et al., 2004).

Table 2.1. Concentration of selenium in raw materials and various wastes (adapted after Lemly, 2004).

Material or waste	Se concentration
Earth's crust	0.2 µg g^{-1}
Surface waters	0.2 µg L^{-1}
Coal	0.4-24 µg g^{-1}
Coal (fly ash)	73-440 µg g^{-1}
Ash settling ponds	87-2700 µg L^{-1}
Fly ash leachate	40-610 µg L^{-1}
Crude oil	500-2200 µg L^{-1}
Crude shale oils	92-540 µg L^{-1}
Refined oils	5-258 µg L^{-1}
Copper ore	20-82 µg g^{-1}
Phosphate	2-20 µg g^{-1}
Municipal landfills	5-50 g L^{-1}

The impact of Se oxyanions and Se(0) on bivalve mollusks and trophic networks was investigated extensively in the San Francisco Bay area (Schlekat et al., 2000; Purkerson et al., 2003; Presser & Luoma, 2006; USEPA, 2010). To complicate matters further, biogenic Se(0) exhibits colloidal properties that make its separation from aqueous solution problematic (Buchs et al., 2013; Staicu et al., 2015b). The most reduced valence states, selenides, Se[-I] and Se[-II], are present under strongly reducing conditions (Fernandez-Martinez & Charlet, 2009).

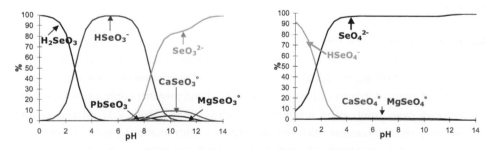

Figure 2.1. Distribution of Se oxyanions as a function of pH in the presence of cations. Conditions used: Pb^{2+}, Ni^{2+}, Cd^{2+}, Zn^{2+} (10^{-6} mol L^{-1}); Ca^{2+}, Mg^{2+} (10^{-4} mol L^{-1}) (Seby, 1999).

2.5. Legislation

The discharge limits for selenium are based on toxicity tests that differ widely between regulatory agencies. This in turn creates disparities among regulations and imposed thresholds (Luoma & Presser, 2009). Selenium is a freshwater priority pollutant under the existing US National Recommended Water Quality Criteria with a Criterion Continuous Concentration or chronic level of 5 μg L^{-1} (USEPA, 2013). In Canada, the target guideline for total selenium content in surface waters was set to 1 μg L^{-1} (Canadian Council of Ministers of the Environment, 2007). In addition to freshwater, British Columbia also developed a Water Quality Guideline for saline water, i.e. 2 μg L^{-1} (Nagpal & Howell, 2001). In Europe, selenium is not listed in the European Commission's Dangerous Substances Directive or the Environmental Quality Standards Directive (Environmental Quality Standard Directive 2008/105/EC).

The presence of selenium in drinking water is also regulated, but the permissible limits vary by as much as 10 times between issuing agencies. Lower values, 10 μg L^{-1}, are required by the European Union, China and Canada, whereas the US impose a relatively higher criterion, 50 μg L^{-1} (European Drinking Water Directive, 1998; Canadian Drinking Water, 2012; USEPA, 2013). For countries lacking legislative framework for drinking water pollutants, the World Health Organization (WHO) provides guidelines on the standards to be achieved. In the case of selenium, the WHO proposes a 40 μg L^{-1} value (WHO, 2011).

2.6. Treatment technologies

2.6.1. Physical treatment

Physical methods used for selenium oxyanions removal include membrane filtration, ion exchange and evaporation.

Membrane filtration

Reverse Osmosis

Reverse Osmosis (RO) uses semipermeable membranes that provide a physical barrier to the passage of particles and soluble chemical species. RO has been employed at pilot-scale and full-scale to effectively treat mining wastewater and agricultural drainage (NAMC, 2010). Mine water produced at Barrick Richmond Hill Mine (South Dakota, US) was treated by RO after an iron reduction and precipitation pre-treatment stage. The final selenium content was reduced from 22 μg L^{-1} to 2 μg L^{-1} with the brine being recirculated to the iron treatment step (Sobolewski, 2005). An emergency RO treatment system was implemented to treat mine water stored in an impoundment from a former gold mine in California

(Golder, 2009). The treatment was effective by reducing 60 μg L^{-1} selenium to < 5 μg L^{-1} within a 4-month operation time. A full-scale hybrid-application (RO + biological treatment) was designed to reduce selenium (30 μg L^{-1}) and Total Dissolved Solids (TDS) (5,000 to 8,000 mg L^{-1}) from a Western US metal mine waste rock leachate to be discharged to surface waters (Gusek et al., 2008). The selenium-enriched reject stream produced by RO was further subjected to an anaerobic bioreactor with the goal of dropping the selenium level below 10 μg L^{-1}. On a pilot scale basis, RO has been tested for agricultural drainage water in California (Red Rock Ranch) having 760 μg L^{-1} Se in the influent and only 1 μg L^{-1} in the effluent (USBR, 2008).

RO, while being effective in selenium reduction in the final effluent (permeate), the reject stream will accumulate and concentrate the contaminants of interest therefore requiring further treatment. Physical, chemical and biological treatments and deep-injection are employed to further treating or disposing of the reject (NAMC, 2010). An evaporator crystallizer system is currently used as a post-treatment stage for membrane filtration reject streams. The end product can be landfilled provided it will meet the disposal requirements of the site. Paint Filter Testing (PFT), Bearing Capacity Estimate and Toxicity Characteristic Leaching Procedure (TCLP) testing are required for landfilling of the waste (NAMC, 2010).

Nanofiltration

Nanofiltration (NF) operates on the same principle like reverse osmosis, but the membrane is charged and has a hydrophobic reject layer that allows it to selectively retain monovalent salts while rejecting divalent salts (Binnie & Kimber, 2009). A marked difference between the two is the amount of pressure put to achieve the same separation efficiency, NF requiring one third of the pressure needed by RO (Sobolewski, 2013). This, together with its more selective nature, makes NF desirable from a financial standpoint.

A laboratory-scale system was used to treat selenium-containing agricultural drainage water (Kharaka et al., 1996). The authors report over 95% removal efficiency from a 1,000 μg L^{-1} initial Se concentration. This lab study was followed by a pilot study in the Imperial Valley of California. The Se influent concentrations were decreased from 42-63 μg L^{-1} to 1.0-3.2 μg L^{-1} (USBR, 2002). Another laboratory-scale study treating uranium mill wastewater by iron coagulation followed by NF reported a 95% Se reduction at pH 10 (Chellam & Clifford, 2002). One of the main limitations of using membrane filtration (RO and NF) is posed by the scaling issue. Hard waters containing Ca^{2+} and Mg^{2+} form salts, thereby damaging or impairing the function of the membrane. In addition, other parameters should be considered, namely, TDS, Total Suspended Solids (TSS), alkalinity, Total Organic Carbon (TOC), sulfate, silica, chloride, iron and the bacterial-algal fouling potential. To overcome these aspects, a filtration pre-treatment stage is necessary but this will add 10-20% to the cost of treatment

(Sobolewski, 2013). If proper pre-treatment is implemented, membrane life is expected to be 2-3 years (Sobolewski, 2013).

Evaporation

Even if evaporation works by a different principle than membrane filtration, they are similar in their capacity to concentrate selenium oxyanions in a smaller water volume.

Evaporation ponds

Evaporation ponds (EP) are used in arid regions where the annual potential evaporation exceeds the annual precipitation, thus creating a water deficit (NAMC, 2010). Solar radiation is used as the main energy source and wastewater is stored in lined ponds. Some reports show selenate reduction to selenite, followed by selenite absorption onto minerals and sediments (NSMP, 2007). EP has been employed to treat selenium-laden agricultural drainage water in Tulare Lake Drainage District (California, US), but this approach achieved only limited success (around 25% efficiency in Se reduction) while the final selenium concentration (15 µg L^{-1}) was still above the discharge limits (NSMP, 2007). A study aimed at the investigation of selenium sinks showed its preferential accumulation in the sediments of an EP (Gao et al., 2007). Other drawbacks of using EP are related to the potential ecological hazard to aquatic birds and the limited use under colder and wet climates. In addition, groundwater contamination events can occur and the generated waste is considered hazardous, requiring special disposal (NSMP, 2007).

Enhanced evaporation systems (EES)

Enhanced evaporation systems provide an alternative to EP, producing concentrated brines by mechanically spraying water in the air using a blower (NAMC, 2010). A higher efficiency, 67%, than EP has been reported (Salton Sea Restoration Program, 2005). The main limitations are due to the scale formation on mechanical parts, energy consumption and drift and air emissions that can be carried away from the treatment area.

Mechanical Evaporator / Crystallizer

Brine concentrators are a subtype of evaporators using a falling-film evaporator device. They use a calcium sulfate seed crystal added in the brine that acts as a nucleation promoter. The brine concentrator can achieve 80-90% water reduction for a concentration factor between 5 and 10 (USEPA, 2009). This technology is sometimes used in concert with other treatment methods, as a post-treatment stage aimed to further reduce the final treated volume.

While efficient, these systems are mechanically- and thermodynamically-complex and they are also expensive. In addition, they require energy and maintenance, plus steam and cooling water (NAMC, 2010). Scale and corrosion are important issues to be considered, especially when dealing with high-strength wastewaters. Pre-treatment is often necessary, adding to the total operating costs. Nevertheless, mechanical evaporators and crystallizers can become important in the context of the Zero Liquid Discharge (ZLD) strategy. ZLD attempts at reducing significantly the outflow stream of industrial processes by returning the recovered water back to the technological flux and by producing a solid residue that could be landfilled. As the legislation for discharging wastewater will become more stringent, the ZLD could considerably gain in importance.

2.6.2. Chemical treatment

Ion Exchange

Ion exchange is a method used to bind undesirable ions present in water to an immobile solid particle which releases desirable ions. The immobile solid particle is a granular medium having a natural (e.g. inorganic zeolites) or a synthetic (e.g. organic resins) origin. Typical ions that bind to ion exchanging resins are H^+ and OH^-, monovalent cations (e.g. Na^+, K^+), divalent cations (e.g. Ca^{2+}, Mg^{2+}), divalent anions (e.g. SO_4^{2-}, SeO_4^{2-}, PO_4^{3-}), charged organic acids, bases, and ionizable biomolecules. Natural and artificial zeolites have shown limited efficiency for selenium treatment, therefore organic strong base anion exchange resins are extensively used today for their higher affinity to SeO_3^{2-} and SeO_4^{2-} (Nishimura & Hashimoto, 2007; Sobolewski, 2013).

A laboratory-scale study investigated various types of resins treating selenium-containing oil refining stripped sour water (Montgomery Watson, 1995). This wastewater contains hydrogen sulfide, ammonia, phenols and chlorides and is of particular concern due to its high corrosiveness to stainless steel components. Amongst the resins tested, the strong base anion resin showed the greatest performance by reducing 4,870 µg L^{-1} selenite down to below 50 µg L^{-1} (Montgomery Watson, 1995). Another study pilot-tested mining water using a silica polyamine resin made from polyethyleneamine impregnated with zirconium. The treated wastewater was acidic (pH 4) and contained both selenium (930 µg L^{-1}) and sulfate (80 mg L^{-1}). Very high removal rates were achieved, with the effluent containing Se concentrations below 1 µg L^{-1} (Golder, 2009).

The competing action of sulfate with selenate plays a role on weakly basic ion exchange resins (Nishimura & Hashimoto, 2007). Sulfate is a competing anion with higher affinity for exchange active sites leading to exhaustion of the resin. In order to overcome this issue, BaCl$_2$ was used to precipitate sulfate. This pretreatment step was then followed by ion exchange and achieved a final selenate level of 100 µg L^{-1} from the initial 1,000 µg L^{-1} load (Golder, 2009).

In order to prevent issues associated with competing oxyanions that will consume the exchange capacity of the resin, pretreatment is required. Other factors to be taken into consideration are related to TSS plugging, organics fouling, high temperatures and the presence of strong oxidants that will negatively impact the performance and shelf-life of the resin (Sobolewski, 2013). In addition to these, the pH of the wastewater must be adjusted before treatment because the exchange capacity of the resin varies as a function of pH. When the ion exchange sites are saturated, the resin can be regenerated using a sodium hydroxide solution. The regeneration solution being highly loaded with selenate and must be further treated onsite or offsite.

Iron Coprecipitation (Ferrihydrite Adsorption)

Precipitation of ferrihydrite (Fh) followed by the adsorption of selenium oxyanions onto its surface is considered a powerful technology for treating selenium-laden waters being approved by USEPA as a treatment option (Sobolewski, 2013). Ferrihydrite is a hydrous ferric oxyhydroxide mineral (Fe_2O_3 * 0.5 H_2O) used in water purification and wastewater treatment because of its high surface to volume ratio and because of its ability to adsorb contaminants of concern, e.g. lead, arsenic, and phosphate (Cornell & Schwertmann, 2003). Fh is considered a form of ferric hydroxide, $Fe(OH)_3$, but being metastable it will transform into more crystalline ferrioxyhydrites (e.g. goethite, lepidocrocite) and ferrioxides (e.g. hematite) (Twidwell et al., 2000). In water treatment, ferrihydrite is formed in-situ by the dissolution of ferric chloride or ferric sulfate, followed by pH adjustment, vigorous stirring and addition of polymers and coagulants (NAMC, 2010). This step is followed by selenium oxyanions adsorption onto the ferrihydrite surface.

The adsorption capacity varies as a function of the wastewater pH and the oxidation state of selenium. The best pH range for effective selenite removal is between 4 to 6 (85-90% removal), the efficiency decreasing with increasing pH, 80-85% at pH 7, and 20-40% at higher pH values (Twidwell et al., 2000). Selenite is strongly adsorbed onto Fh, while selenate is only loosely bound onto Fh and therefore this method cannot be used effectively for selenate removal (Twidwell et al., 2000). The reason behind this difference in adsorption has a mechanistic explanation. While selenite is adsorbed to the ferrihydrite matrix through an inner-sphere complex, selenate forms an outer-sphere complex and can be more easily replaced by other ions present in solution (Hayes et al., 1987).

Merrill et al. (1986) performed an early study on selenium treatment using iron coprecipitation. They treated a selenium-laden wastewater containing 40-60 µg L^{-1} selenite down to below 10 µg L^{-1} using 14 mg L^{-1} iron at an optimum pH of 6.5, generating 2.1-3.1 kg sludge kg^{-1} of iron added. The action of competing anions at pH 7 was reported in the following order: phosphate > silicate >As(V) > bicarbonate/carbonate > Se(IV) > oxalate > fluoride > Se(VI) > sulfate (Balistrieri & Chao, 1990). In the case of a complex wastewater containing multiple chemical species, pre-treatment is required. Another important issue is related to the metastability of ferrihydrite over time which can mature to the more thermodynamically stable goethite or hematite accompanied by a large decrease in surface area and the potential release of co-precipitated contaminants including selenium (Twidwell et al., 2000).

Ferrous Hydroxide

Ferrous hydroxide, $Fe(OH)_2$, is used to reduce selenate to selenite, which is subsequently adsorbed on ferrihydrite monohydrate amorphous solids. $Fe(OH)_2$ is precipitated at neutral pH by the addition of NaOH to $FeSO_4$ (Lalvani, 2004). The reduction of selenate and the adsorption of selenite are best accomplished under reducing conditions at a pH of 8-9 (Twidwell et al., 2000).

In a Flue Gas Desulfurization (FGD) wastewater treatment trial, $FeCl_2$ was used for the generation of $Fe(OH)_2$ at pH of 8. Polymers were added to the effluent and allowed it to settle in a clarifier for 3 days. Next, the overflow from the clarifier was passed through a media filter. A treatment efficiency of 84% has been reported, but the final total Se concentration was still 55 μg L^{-1}. Care must be taken for the overshooting of the chloride dosed, as the Se removal efficiency drops to around 60% at Cl^- levels of around 15,000 mg L^{-1} (NAMC, 2010).

Zero Valent Iron

Selenium removal using the Zero Valent Iron (ZVI) technology involves the complex redox and adsorption interactions between metallic iron and selenium oxyanions present in the wastewater. Iron acts as a strong oxidant being the electron donor but also as the catalyst for the reduction of oxyanions (Frankenberger et al., 2004). Water solutions containing Dissolved Oxygen (DO) will corrode metallic iron (Fe^0) by forming ferrous and ferric hydroxides. In addition to DO, oxyanions (e.g. NO_3^-, CO_3^{2-}, PO_4^{3-}, SeO_3^{2-}, SeO_4^{2-}) present in the water will also contribute to the overall oxidation (corrosion) of iron.

By mixing $Fe(OH)_2$ and $Fe(OH)_3$ at an optimal pH range of 4-5, green rust is formed. Green rust is a complex ferrihydroxide coprecipitate that has been shown to sequentially reduce selenate to selenite to elemental selenium (NAMC, 2010). ZVI has also the potential to directly reduce Se(VI) and Se(IV) to Se(0) (Eq. 2.1 and Eq.2.2). Alternatively, selenite can be adsorbed onto ferrihydrite and ferri-hydroxide amorphous solids formed as byproducts of ZVI oxidation.

$$3Fe^0_{(s)} + SeO_4^{2-} + 8H^+ \rightarrow 3Fe^{2+} + Se^0 + 4H_2O \hspace{3cm} (2.1)$$

$$2Fe^0_{(s)} + SeO_3^{2-} + 6H^+ \rightarrow 2Fe^{2+} + Se^0 + 3H_2O \hspace{3cm} (2.2)$$

For a higher efficiency, ZVI can be employed in concert with other treatment techniques. To treat mine wastewater, ZVI was used to reduce selenate to selenite, followed by the reduction to elemental selenium using iron co-precipitation. Because the initial 100 µg L^{-1} Se was reduced to only 12-22 µg L^{-1}, RO was implemented as a polishing step prior to effluent discharge (Sobolewski, 2005). ZVI treatment can also be coupled to biological (post)treatment (Zhang Y. et al., 2005), which is effective in treating inorganic and organic selenium (see Biological treatment, section 2.6.3).

Flue gas desulphurization wastewater generated in a coal-fired power plant was treated by metallic iron powder with a significant Se reduction: from 7,270 µg L^{-1} to 159 µg L^{-1} (EPRI, 2009). In the first stage, ferrous hydroxide was precipitated by raising the pH and nitrate was chemically reduced to ammonia.

Another study investigated the treatment of Se-laden stripped sour water containing mainly selenocyanate, $SeCN^-$, in the 250-500 µg L^{-1} range and yielded 79% reduction (Shamas et al., 2009). Selenocyanate is another water-soluble reduced form of Se of that raises challenges in terms of toxicity and removal from wastewaters (Manceau & Gallup, 1997). It is mainly produced by petroleum refining and mining Industries (Meng et al., 2002). In addition, SeCN- is an important pollutant in mining wastewaters where cyanides leaches Se-containing ores (de Souza et al., 2002). Currently, the most used method to tread $SeCN^-$ is by oxidation to SeO_3^{2-} and they further treatment (e.g. precipitation) of this oxyanion (NAMC, 2010).

A mining wastewater containing selenate was treated sequentially using elemental iron to reduce hexavalent Se to Se(IV), followed by $Fe_2(SO_4)_3$ addition and precipitation at pH 4.5, a reaction catalysed by $CuSO_4$. Even if the treatment process was optimized, the final Se effluent concentrations (12-22 µg L^{-1}) were still above the discharge limits. In order to decrease the effluent Se concentration further, the treatment was complemented by a reverse osmosis unit (Sobolewski, 2013).

The complexity of the wastewater matrix will negatively impact the efficiency of ZVI treatment. Competing oxyanions and DO, narrow pH range (4-7), and temperature slow down the reaction kinetics. In addition, metals and suspended solids will coat the surface of iron particles, leading to passivation and decreased treatment efficiency. Therefore, pre-treatment, as well as post-treatment for proper sludge disposal, are mandatory (NAMC, 2010).

Catalyzed reduction / Cementation

Catalyzed cementation is a ZVI treatment variation developed at Montana Tech (US) wherein copper or nickel addition is used to enhance the electrochemical potential and thus improve the reduction of the selenium oxyanions by the elemental iron (MSE, 2001; Sobolewski, 2013). Catalyzed cementation can treat both selenium oxyanions. An USEPA sponsored demonstration project was implemented at the Kennecott Utah Copper Corporation site to treat high selenate-laden wastewater (around 1,600 µg L^{-1}). The treatment was effective in producing low level Se effluent (3 µg L^{-1}) (Golder, 2009).

As a general conclusion of the iron-based sorbents, the presence of DO, nitrate and bicarbonate limits the performance of the treatment. Residuals handling and disposal constitute additional drawbacks. Another major issue is linked to the potential formation of toxic H_2Se.

Electrochemical reduction

Baek et al. (2013) investigated the removal of selenate from synthetic water using iron and mixed metal oxide (titanium coated with IrO_2 and Ta_2O_5) electrodes. They found that selenate removal was not directly proportional to the applied current but dependent on the concentration of $Fe(OH)_2$, proposed as the reducing agent for selenate and selenite. Ferrous oxide undergone oxidation under the action of DO present in solution, whereas SeO_4^{2-} was reduced sequentially to SeO_3^{2-} and then to Se(0) or Se(-II). Even if the removal efficiencies were high, the residual selenium (0.79 mg L^{-1}) at the end of the treatment was still high above the discharge limit. No investigation on the electrochemical removal of selenite was reported so far.

Coagulation-flocculation

Colloidal biogenic Se(0) can undergo solid-liquid separation by the addition of metallic salts. Metallic salts hydrolyze spontaneously with the formation of a series of metastable hydrolysis products that transit towards metal hydroxides (Richens, 1997). Because colloids are held in suspension due to the electrostatic repulsion forces, the presence of counter ions brings about neutralization of the electric charge and diminishes their colloidal stability (Khandegar & Saroha, 2013). Destabilized colloidal particles are adsorbed onto metal (oxy)hydroxides, followed by precipitation (Hanai & Hasar, 2011). As a consequence of these mechanisms, the colloids aggregate and settle down. Staicu et al. (2015b) showed that aluminum sulfate is an effective coagulating agent against colloidal Se(0) produced by a mixed microbial culture.

Electrocoagulation

Electrocoagulation (EC) can be used as an alternative to chemical coagulation dosing for the removal of colloidal Se(0) from wastewater. It differs from the latter in the continuous electrogeneration of the coagulating agent by the passing of an electric current through an electrolytic cell containing metal sacrificial electrodes (e.g. iron, aluminum) (Figure 2.2A). The electrogenerated coagulants (e.g. Al^{3+}, $Al(OH)_3$, Fe^{2+}/Fe^{3+}, $Fe(OH)_3$) react with the dissolved or particulate pollutants (Figure 2.2B) leading to their sedimentation and separation from the treated solution (Mollah et al., 2004).

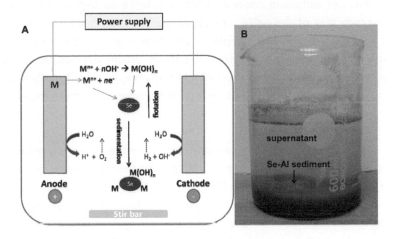

Figure 2.2. (A) Electrocoagulation setup. Notes: M = metal (e.g. Al, Fe) and (B) Colloidal Se(0) treated by electrocoagulation with aluminum electrodes (Staicu et al., 2015a).

Electrocoagulation of colloidal Se(0) produced by a strain of *Pseudomonas moraviensis* has been shown to be effective using iron and aluminum electrodes (Staicu et al., 2015a). The best colloidal Se(0) turbidity removal (97%) was achieved using iron electrodes at 200 mA. Aluminum electrodes generated 96% removal at a slightly higher current intensity (300 mA). As a consequence of the mineralogical state of the sediments, the Se-Al sediment was three times more voluminous than the Se-Fe sediment. The TCLP test showed that the Fe-Se sediment released Se below the regulatory level (1 mg L^{-1}), whereas the Se concentration leached from the Al-Se sediment was 20 times in excess. This is particularly important in view of the management and safe disposal of the generated sediments.

2.6.3. Biological treatment

Microbial conversion of selenium oxyanions

Even if dissimilatory reduction of selenium oxyanions has been described (Oremland et al., 1989; Lovely, 1993), the general picture of bacterial selenium metabolism is still inconclusive (Stolz & Oremland, 1999). Aerobic reduction of selenite appears to be ubiquitous amongst phylogenetically-diverse bacterial groups, as they probably share common metabolic pathways used for the reduction of other compounds, as nitrate or sulfate (Sura-de Jong et al., 2014). Staicu et al. (2015c, unpublished results) described a novel strain of *Pseudomonas moraviensis* that can withstand concentrations of selenite as high as 120 mM, showing growth and production of red Se(0). The conversion of SeO_3^{2-} under aerobic conditions occurred at a high rate, 0.27 mM h^{-1}, depleting 10 mM of selenite within 48 h of incubation. In comparison, the *Shewanella oneidensis* MR-1 strain grown anaerobically at 30°C, in LB media amended with 0.5 mM selenite and 20 mM fumarate, had significantly lower selenite reduction rates (between 0.5 and 1.5 µM h^{-1}) for the wildtype and three mutants (Li et al., 2014).

In the case of the aerobic reduction of selenate, only a handful of species/strains have been described to date (reviewed in Kuroda et al., 2011). This limitation might be caused by the constitutive lack of a selenate reductase (Schröder, 1997; Watts et al., 2003). Several studies used mixed culture inocula (anaerobic sludge) but their focus was not on the identification of the selenium-reducing bacteria but on the process optimization of soluble selenium removal (Lenz et al., 2008; Soda et al., 2011; Hageman et al., 2013).

Compared to the pure microbial cultures, the main advantage of using mixed cultures (consortia) stems from the fact that the interactions within the consortia provide enhanced metabolic capabilities and tolerance to environmental stressors like high salinity (Ike et al., 2000). In addition, by the nature of wastewaters and the bioreactor design, pure cultures will be contaminated and potentially outcompeted by other bacteria (Lenz et al., 2006). The reduction of Se oxyanions using anaerobic granular sludge leads to the formation of red Se(0) that sticks to the outer surface of the sludge granule (Figure 2.3).

Granular Sludge Bioreactors

Upflow Anaerobic Sludge Blanket (UASB) reactors have been pilot-tested for selenium removal at the Adams Avenue Agricultural Drainage Research Center in San Joaquin Valley (California). The influent had a total Se content of 500 µg L^{-1} and the removal efficiency ranged from 58 to 90% (NAMC, 2010). Amongst the limitations encountered were the long acclimation period of the granular sludge (around 6 months), the short-circuiting of the bioreactor caused by the accumulation of gas in the sludge, and the big variability in selenium removal due to the temperature sensitivity of the process. When the temperature dropped from 15 °C to 7 °C, the removal efficiency decreased from 88% to 35%, prompting the insulation of the tanks during the cold season.

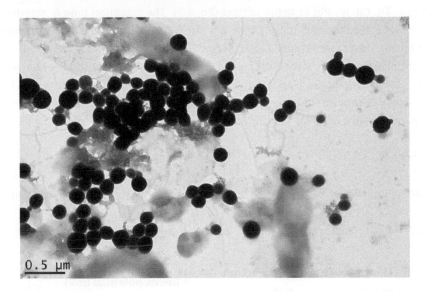

Figure 2.3. Biogenic selenium nanoparticles produced by anaerobic granular sludge (Staicu et al., 2015b).

Fluidized Bed Reactor

Fluidized Bed Reactor (FBR) technology uses a fixed-film that is attached to fine sand and activated carbon media. The technology was patented by Envirogen Technologies. Treated wastewater is transferred to a downstream liquid/solids separation system where the biological solids and elemental selenium are separated. Coal mining wastewaters containing Se in the 22-500 µg L^{-1} range were treated below 10 µg L^{-1} (Envirogen, 2011).

Attached Growth Downflow Filter: ABMet® System

The ABMet® (*Advanced Biological Metals Removal*) System employs anaerobic bacteria fixed to a granular activated carbon (GAC) bed (Pickett et al., 2008) and, in contrast to the UASB technology, the wastewaters flow downwards. ABMet® uses mesophilic bacteria (13-38 °C) and consists of two bioreactors in series in multiple trains with a typical 4 to 8 hours detention time (Picket et al., 2008). The wastewater entering the system has a positive OPR (+200/+300 mV) and is pumped downflow through the GAC bed. The molasses-based carbon source injected in different compartments of the bioreactor creates negative ORP zones, required for different types of reactions (Table 2.2). Molasses is a cheap source of carbon and iron required for bacterial metabolism (NAMC, 2010). According to the vendor's technical reports (Pickett et al., 2008), a sequential cleaning up of the wastewater takes place, starting with denitrification, followed by selenite and selenate removal. Below -200 mV, sulfate-reducing bacteria (SRB) form sulfides (S^{2-}) involved in the precipitation of trace metals (e.g. Zn, Cu, Ni, Pb) as metal sulfides (Eq. 3.3). The retention time is determined by the wastewater type treated, ranging from 6 h (for Flue Gas Desulfurization wastewater) to 3 h (for mining wastewater) (Joel Citulski, personal communication, October 2014). In order to prevent plugging, the reactor must be degassed (on a daily basis) and backwashed (on a monthly basis) periodically (Joel Citulski, personal communication, October 2014). The effluent water generated is further aerated in a polishing step to increase the DO level and reduce excess soluble organics.

$$M^{n+} + S^{2-} \rightarrow MS_{(s)} \tag{3.3}$$

where M^{n+}, metal in the oxidation state n; S^{2-}, sulfide; $MS_{(s)}$, metal sulfide in solid state.

Table 2.2. ABMet$^{\circledR}$ reactor gradational ORP zones (after Picket et al., 2008).

Flow through bioreactor	Final electron acceptor	Redox potential (mV)
	O_2	> 0
	NO_3^-	< 0
	NO_2^-	< -50
	SeO_4^{2-}	< -100
	SeO_3^{2-}	< -150
	SO_4^{2-}	< -200

Both pilot and full-scale ABMet$^{\circledR}$ systems have been tested for a variety of selenium influenced wastewater types, e.g. issued by power generation, oil and gas, mining, agricultural sectors. A pilot test demonstration was performed at the Kennecott Utah Copper Corporation. The process reduced 1,950 µg L^{-1} Se to below 2 µg L^{-1} (MSE, 2001). Another pilot test managed to reduce 43 µg L^{-1} Se to below 5 µg L^{-1} at a coal mine in West Virginia (NAMC, 2010). Agricultural Se-influenced water from Panoche Drainage District (California) was treated for selenium and nitrate, both contaminants being brought below 5 µg L^{-1} and 5 mg L^{-1}, respectively (NAMC, 2010). At a full-scale level, a system was tested at a gold mine in South Dakota for the treatment of both nitrate and selenium. 30 mg L^{-1} NO_3^- was reduced to 10 mg L^{-1} in a four-cell configuration, whereas 2 cells were used to reduce 100 µg L^{-1} Se to 2 µg L^{-1} (Maniatis & Adams, 2003). Currently, 8 full-scale ABMet$^{\circledR}$ systems are in operation in North America (USA and Canada) and Europe (Belgium): 5 treating FGD, and 3 treating coal mining, agricultural runoff and precious metal recycling wastewaters (Joel Citulski, personal communication, October 2014).

Electro-Biochemical Reactor

The electro-biochemical reactor (EBR) produced by INOTEC Inc. provides electrons to the microbial community using graphite electrodes and a low voltage potential (1-3 V) (Opara et al., 2014). Certain bacterial consortia can transfer electrons from one species to another (Direct Interspecies Electron Transfer, DIET) or accept electrons produced by electrodes that are used to reduce organic/inorganic compounds (Gregory et al., 2004; Lovely, 2008). In addition to replacing the electron donor, the electrodes ensure a better way to control the ORP needed to sequentially remove nitrate, Se oxyanions and sulfate. A laboratory bench- and on-site pilot-scale study reported the decrease of Se concentrations (35-531 µg L^{-1}) from mining wastewaters below discharge targets (5-10 µg L^{-1}). Furthermore, nitrate (11-170 mg L^{-1}) was reduced to below 1 mg L^{-1}, the efficiency increasing as a function of the retention time (from 6 to 18 h) (Opara et al., 2014).

Hydrogen-Based Membrane Biofilm Reactor

Membrane biofilm reactors (MBfR) have been tested for the removal of selenium oxyanions using synthetic wastewaters (Nerenberg & Rittmann, 2004; Chung et al., 2006). Hydrogen is used as the electron donor that is delivered from inside the membrane lumen to the biofilm growing on the membrane surface. The Se removal efficiency reported was around 95% (from 260 µg L^{-1} to 12 µg L^{-1}) (Chung et al., 2006). Further studies have reported a maximum selenate removal flux of 362 mg Se $m^{-2} d^{-1}$ when treating real FGD wastewater (van Ginkel et al., 2011a). It is interesting to note that the presence of nitrate and nitrite did not compete with the removal of selenate but, on the downside, NO_x were not denitrified which entails a further denitrification stage needed as a polishing post-treatment step (van Ginkel et al., 2011b). A microbial community study conducted recently showed that excess nitrate (surface loading larger than 1.14 g of N $m^{-2} d^{-1}$) had an inhibitory effect on SeO_4^{2-} reduction and *Hydrogenophagasp.*, an autotrophic denitrifier β-proteobacterium, was positively correlated with the NO_3^- flux (Lai et al., 2014). This finding suggests the need to remove excess nitrate prior to selenate treatment in order to avoid competition.

Major drawbacks of this technology are: the delivery of H_2 to the biofilm, insufficient reproducibility and the time needed for the proliferation of Se-reducing microbial community. While hydrogen is a cheaper electron donor than its carbon-based counterpart, special care should be taken as hydrogen is an explosive gas.

Passive biological treatment

Passive treatment systems are designed as a 'walk-away' solution, without an operator and limited or no maintenance (Sobolewski, 2010). The Biopass system was initially developed for passively treating cyanide heap leachate resulting from gold extraction (Cellan et al.,

1997). The system is built in an excavated area isolated with a geomembrane and the substrate layer (gravel, composted manure, wood chips, alfalfa or peat) is filled in. The wastewater is seeping gravitationally through the decaying organic matter and treated under anaerobic conditions by the microbial activity within the bioreactor and collected at the bottom using an underdrain device (NAMC, 2007).

A full-scale bioreactor at a gold mine in Montana reduced influent Se concentrations form 20 µg L^{-1} to less than 1 µg L^{-1} (Golder, 2009). An 18-month pilot study conducted at the Brewer Mine (South Carolina, US) of mine-influenced water derived from an acidic pit lake reported 97% Se removal (from 1,500 µg L^{-1} to 50 µg L^{-1}) (Golder, 2009).

While an interesting technology, the treatment potential of passive reactor systems is limited because the organic substrate will be depleted by the microbial communities over time and thus requires a period replacement. Another potential issue is the resolubilization of the immobilised Se (Knotek-Smith et al., 2006).

Constructed Wetlands

Engineered wetlands are ecosystems designed and constructed for the treatment of large volumes of wastewater (Table 2.3). They use vegetation, soil and their associated microbial activity to convert Se oxyanions into Se(0) and selenides. The anoxic environment created by the decomposition of wastewater and biological detritus favors the anaerobic reduction of Se oxyanions. Se(0) generated through microbial activity is bound to the wetland sediments (Oremland, 1993). Unlike natural wetlands, engineered wetlands contain monocultures of cattails (*Typha* spp.) or bulrushes (*Schoenplectus* spp.) and have regulated inflow rates and water depths (NAMC, 2010). The major players involved in the transformation of Se are bacteria, fungi and algae. Microbial volatilization of Se has been reported by Hansen et al. (1998) and Frankenberger et al. (2004) and this can be an important route of Se sink. Various configurations of constructed wetlands have been proposed, the main designs being surface flow and subsurface flow wetlands.

The mass balance of Se within the wetland is of paramount importance. Gao et al. (2003) found that sediments are an important (56%) sink of Se that enters in the system. The other fractions were: 3% assimilated by vegetation, 2% volatilized, 4% infiltrated into the groundwater and 35% lost in the effluent. Se was reduced from 20 µg L^{-1} to 3-6 µg L^{-1} (up to 85% removal) at a residence time of 7 days and temperature has been shown to be a rate-limiting factor below 10-15 °C (Gao et al., 2003).

A microcosm study using *Thalia*, cattails and rabbit-foot (*Polypogon monspeliensis*) grasses attempted to treat wastewaters containing selenocyanate (Ye et al., 2003). The system achieved a 64% removal rate by decreasing Se from 1440 µg L^{-1} to 510 µg L^{-1}, while the primary sink of Se (63%) was within the sediments, 18% in the effluent, and 11% in the surface water. Only small fractions were assimilated by the plants (4%) or volatilized (3%).

Table 2.3. Selenium removal in full-scale surface flow wetlands (Adapted from Kadlec & Wallace, 2009; NAMC, 2010).

Site	Waste(water)	Influent $Se_{tot}(\mu g\ L^{-1})$	Effluent $Se_{tot}(\mu g\ L^{-1})$	Removal efficiency (%)
Great Falls (Montana)	Drainage	26	1	96
Richmont (California)	Refinery wastewater	25	5	80
Albright (Pennsylvania)	Coal ash leachate	4	2	50
Corcoran (California)	Agricultural drainage	16	9	44
Imperial (California)	Agricultural drainage	7.1	5.9	17
PG-4 (Confidential)	FGD	170	150	12

Some major limitations of using wetlands to treat Se-laden wastewaters are related to the exposure potential to ecological receptors, large footprints (determined by the low Hydraulic Retention Time - HRT - of several days to one week), the clogging of the system over time that will affect the hydraulic conductivity of the media, climate conditions (temperature and precipitations), requirement of post-treatment (aeration) to bring the dissolved oxygen at a suitable level prior to discharge into surface waters, long term management requirements, potential of contaminated water percolating through the degraded liner (geomembrane) and seeping into groundwater.

Enhanced in situ microbial reduction

Enhanced *in situ* microbial reduction uses microbial specialists, nutrients and organic amendments which are added to Se-contaminated water, e.g. pit lakes, groundwater influenced from mining and mine workings. Organic amendments are added to deplete the dissolved oxygen and therefore create an anoxic environment for the growth of selenium reducing anaerobic bacteria (Nelson et al., 2003).

This technique has been implemented at full-scale at a New Mexico Reclamation Project and brought the Se concentration down to less than 5 $\mu g\ L^{-1}$ (from initial 50-100 $\mu g\ L^{-1}$) (Sobolewski, 2005). At Sweetwater Pit Lake (Wyoming), sugar, alcohols, fats, proteins, phosphate and nitrate have been added over a 2-month period to an open pit uranium mine, which resulted in the depletion of 460 µg L^{-1} Se to less than 10 µg L^{-1} (Nelson et al., 2003). The addition of phosphate induced algal bloom events which provided an organic carbon source for Se microbial reduction. Pit lake treatment was also performed at Beal Mountain Mine (Montana) using organic carbon. Se levels were decreased from 45 µg L^{-1} to 2-3 µg L^{-1} (Nelson et al., 2003).

Because nutrient addition creates anoxic conditions, this approach is not suitable for environments containing aquatic life. Water treatment by enhanced *in situ* microbial reduction requires post-treatment prior to discharge in surface waters to increase the dissolved oxygen levels, filtrate particulate metals and Se(0), and remove hydrogen sulphide (Nelson et al., 2003). Moreover, the major drawbacks are related to the suitability of the microbial inoculum used since some pit lake waters are saline, cold and contain oxygen levels prohibitive to the proliferation of Se anaerobic microbial specialists.

2.7. Conclusions and Outlook

Selenium is present in a variety of wastewaters resulting from petrochemical (coal, gas, crude oil), metal refining or agricultural (Se rich and seleniferous soils) activities. The main challenges encountered in the treatment of selenium-laden wastewaters are:

- the presence of Se in several chemical forms (SeO_3^{2-}, SeO_4^{2-}, Se(0), SeCN$^-$); but also chemical species such as nitrate and sulfate hampering the efficiency of biological processes;
- removal is limited by the feasible ranges of design flows;
- the presence of Se in relatively low concentrations;
- the wastewater matrix can confound the removal efficiency;
- the treatment generates residual by-products that must be further treated or disposed of safely in order to prevent re-release.
- some wastewaters (mining, agriculture) are produced in very large amounts and their source is diffuse.

Physical-chemical treatment approaches are effective but suffer from high and prohibitive investment and operational costs. On the other hand, the biological treatment is limited in that it requires a relative high footprint and sometimes shows an erratic performance. It can be concluded that there is no best technology for treating Se containing wastewaters, but that a mixed approach including physical-chemical and biological technologies is more suitable to bring Se levels in wastewaters to below the discharge permits.

2.8. References

Alfthan, G., Eurola, M., Ekholm, P., Venäläinen, E.R., Root, T., Korkalainen, K., Hartikainen, H., Salminen, P., Hietaniemi, V., Aspila, P., & Aro, A. (2014). Effects of nationwide addition of selenium to fertilizers on foods, and animal and human health in Finland: From deficiency to optimal selenium status of the population. *J. Trace Elem. Med. Biol.* doi: 10.1016/j.jtemb.2014.04.009.

Amouroux, D., Liss, P.S., Tessier, E., Hamren-Larsson, M., & Donard, O.F.X. (2001). Role of oceans as biogenic sources of selenium. *Earth Planet. Sci. Lett. 189*, 277-283.

Baek, K., Kasem, N., Ciblak, A., Vesper, D., Padilla, I. & Alshawabkeh, A.N. (2013). Electrochemical removal of selenate from aqueous solutions. *Chem. Eng. J. 215-216*, 678-684.

Balistrieri, L.S. & Chao, T.T. (1990). Adsoprtion of selenium by amorphous iron oxyhydroxides and manganese dioxide. *Geochim. Cosmochim. Acta 54*, 739-751.

Binnie, C. & Kimber, M. (2009). Basic Water Treatment. Royal Society of Chemistry, UK.

Breen, B. (2009). Testimony before the U.S. House of Representatives Subcommittee on Water Resources and Environment, April 30 2009. Available from: http://www.epa.gov/ocir/hearings/testimony/111_2009_2010/2009_0430_bb.pdf

Buchs, B., Evangelou, M.W.H., Winkel, L.H.E., & Lenz, M. (2013). Colloidal properties of nanoparticular biogenic selenium govern environmental fate and bioremediation effectiveness. *Environ. Sci. Technol. 47* (5), 2401-2407.

Canadian Council of Ministers of the Environment (2007). Canadian water quality guidelines for the protection of aquatic life: Summary table. In: *Canadian environmental quality guidelines: Canadian Council of Ministers of the Environment*, Winnipeg, SK, Canada.

Cellan, R., Cox, A., Uhle, R., Jenevein, D., Miller, S. & Mudder, T. (1997). Design and construction of an in situ anaerobic biochemical system for passively treating residual cyanide drainage. 1997 National Meeting of the American Society for Surface Mining and Reclamation, Proceedings, Austin, TX.

Chapman, P.M., Adams, W.J., Brooks, M., Delos, C.G., Luoma, S.N., Maher, W.A., Ohlendorf, H.M., Presser, T.S., & Shaw, P. (2010). *Ecological assessment of selenium in the aquatic environments*. SETAC Press, Pensacola, Florida, USA.

Chellam, S. & Clifford, D.A. (2002). Physical-chemical treatment of groundwater contaminated by leachate from surface disposal of uranium tailings. *J. Environ. Eng. 128*, 942-952.

Chung, J., Nerenberg, R. & Rittmann, B.E. (2006). Bio-reduction of selenate a hydrogen-based membrane biofilm reactor. *Environ. Sci. Technol. 40*, 1664-1671.

Cornell, R.M. & Schwertmann, U. (2003). *The iron oxides: structure, properties, reactions, occurrences and uses*. Wiley–VCH, Weinheim, Germany.

Council Directive 98/83/EC of 3 November 1998 on the quality of water intended for human consumption. Off. J. Eur. Union. 330, 32-54. Available from: http://eur-lex.europa.eu/LexUriServ/LexUriServ.do?uri=OJ:L:1998:330:0032:0054:EN:PDF

de Souza, M.P., Pickering, I.J., Walla M. & Terry, N. (2002). Selenium assimilation and volatilization from selenocyanate-treated Indian Mustard and Muskgrass. *Plant Physiol. 128*, 625-633.

Electric Power Research Institute (EPRI) (2009). Selenium removal by iron cementation from a coal-fired power plant flue gas desulfutrization wastewater in a continuous flow system: A pilot study. Palo Alto, CA.

Envirogen Technologies (2011). Treatment of selenium-containing coal mining wastewater with Fluidized Bed Reactor.Available from: http://www.envirogen.com/files/files/ETI_Selenium_GrayPaper_V_FINAL.pdf

Environmental Quality Standard Directive 2008/105/EC. Available from: http://eur-lex.europa.eu/LexUriServ/LexUriServ.do?uri=OJ:L:2008:348:0084:0097:EN:PDF

Fernandez-Martinez, A. & Charlet, L. (2009). Selenium environmental cycling and bioavailability: a structural chemist point of view. *Rev. Environ. Sci. Biotechnol. 8*, 81-110.

Frankenberger Jr., W.T., Amrhein, C., Fan, T.W., Flaschi, D., Kartinen, E., Kovac, K., Lee, E., Ohlendorf, H.M., Owens, L., Terry, N. & Toto, A. (2004). Advanced treatment technologies in the remediation of seleniferous drainage waters and sediments. *Irrig. Drain 18*, 517-522.

Gao, S., Tanji, K.K., Peters, D.W., Lin, Z. & Terry, N. (2003). Selenium removal from irrigation drainage water flowing through constructed wetland cells with special attention to accumulation in sediments. *Water Air Soil Poll. 144*, 263-284.

Gao, S., Tanji, K.K., Dahlgren, R.A., Ryu, J., Herbel, M.J. & Higashi, R.M. (2007). Chemical status of selenium in evaporation basins for disposal of agricultural drainage. *Chemosphere 69*, 585-594

Golder Associates Inc. (Golder) (2009). Literature review of treatment technologies to remove selenium from mining influenced water. Prepared for Tech Coal Limited, Calgary AB. 0842-0034. Available from: http://namc.org/docs/00057713.PDF

Gregory, K.B., Bond, D.R., & Lovley, D.R. (2004). Graphite electrodes as electron donors for anaerobic respiration. *Environ. Microbiol. 6*(6), 596-604.

Guidelines for Canadian Drinking Water Quality (2012). Available from: http://www.hc-sc.gc.ca/ewh-semt/pubs/water-eau/2012-sum_guide-res_recom/index-eng.php

Gusek, J., Conroy, K. & Rutkowski, T. (2008). Past, present and future for treating selenium-impacted water. In: *Tailings and Mine Waste 2008*, Proceedings of the 12th International Conference, CRC Press, Boca Raton, FL, USA.

Hageman, S.P.W., van der Wejiden, R.D., Wejima, J. & Buisman, C.J.N. (2013). Microbiological selenate to selenite conversion for selenium removal. *Water Res. 47*, 2118-2128.

Hamilton, S.J. (2004). Review of selenium toxicity in the aquatic food chain. *Sci. Total. Environ. 326*, 1-31.

Hanai, O. & Hasar, H. (2011). Effect of anions on removing Cu^{2+}, Mn^{2+} and Zn^{2+} in electrocoagulation process using aluminum electrodes. *J. Hazard. Mater. 189*, 572-576.

Hansen, D., Duda, P.J., Zayed, A. & Terry, N. (1998). Selenium removal by constructed wetlands: role of biological volatilization. *Environ. Sci. Technol. 32*, 591-597.

Hartikainen, H. (2005). Biogeochemistry of selenium and its impact on food chain quality and human health. *J. Trace Elem. Med. Biol. 18*(4), 309-318.

Hayes, K.F., Roe, A.L., Brown, G.E., Hodgson, K.O., Leckie, J.O. & Parks, G.A. (1987). In situ X-ray absorption study of surface complexes: Selenium oxyanions on alpha FeOOH. *Science 17*, 139-145.

Haygarth, P.M. (1994). Global importance and global cycling of selenium. In: Frankenberger Jr., W.T., & Benson, S. *Selenium in the Environment*. New York, USA. pp 1-28.

Hodson, P.V., Whittle, D.M. & Hallett, D.J. (1984). Selenium contamination of the Great lakes and its potential effects on aquatic biota. In: Nriagu, J.O., & Simmons, M.S. (eds), *Toxic contaminants in the Great lakes*. Wiley, New York, USA, pp. 317-391.

Ike, M., Takahashim, K., Fujitam, T., Kashiwa, M., & Fujita, M. (2000). Selenate reduction by bacteria isolated from aquatic environment free from selenium contamination. *Water Res. 34*, 3019-3025.

International Energy Agency (IEA) (2013). Medium-Term Coal Market Report 2013 - *Market Trends and Projections to 2018*. ISBN 978-92-64-19120-4.

International Energy Agency (IEA) (2014). Available from: http://www.iea.org/topics/coal/

Kabata-Pendias, A. (2000). *Trace elements in soil and plants*. 3rd Ed, CRC Press, Boca Raton, FL, USA.

Kadlec, R.H. & Wallace, S.D. (2009). *Treatment wetlands*. 2nd Ed, CRC Press, Boca Raton, FL, USA.

Khandegar, V., & Saroha, A.K. (2013). Electrocoagulation for the treatment of textile industry effluent - A review. *J. Environ. Manage. 128*, 949-963.

Kharaka, Y.K., Ambats, G. & Presser, T.S. (1996). Removal of selenium from contaminated agricultural drainage water by nanofiltration membranes. *Appl. Geochem. 11*, 797-802.

Kuroda, M., Notaguchi, E., Sato, A., Yoshioka, M., Hasegawa, A., Kagami, T., Narita, T., Yamashita, M., Sei, K., Soda, S., & Ike, M. (2011). Characterization of *Pseudomonas stutzeri* NT-I capable of removing soluble selenium from the aqueous phase under aerobic conditions. *J. Biosci. Bioeng. 122*, 259-264.

Lai, C.Y., Yang, X., Tang, Y., Rittmann, B.E. & Zhao, H.P. (2014). Nitrate shaped the selenate-reducing microbial community in a hydrogen-based biofilm reactor. *Environ. Sci. Technol. 48*(6), 3395-3402.

Lalvani, S.B. (2004). Selenium removal from agricultural drainage water: Lab scale studies. Final report to the Department of Natural Resources, Sacramento, CA. Available from:

http://www.water.ca.gov/pubs/drainage/selenium_removal_from_agricultural_drai nage_water_lab_scale_studies/seremvl.pdf

Lemly, A.D. (1998). *Selenium assessment in aquatic ecosystems: A guide for hazard evaluation and water quality criteria*. Springer, New York, USA.

Lemly, A.D. (2002). Symptoms and implications of selenium toxicity in fish: the Belews Lake case example. *Aquat. Toxicol. 57*(1-2), 39-49.

Lemly, A.D. (2004). Aquatic selenium pollution is a global environmental safety issue. *Ecotox. Environ. Safe.59*, 44-56.

Lenz, M., Gmerek, A. & Lens, P.N.L. (2006). Selenium speciation in anaerobic granular sludge. *Int. J. Environ. Anal. Chem. 86*, 615-627.

Lenz, M., van Hullebusch, E., Hommes, G., Corvini, P. & Lens, P.N.L. (2008). Selenate removal in methanogenic and sulfate reducing upflow anaerobic sludge bed reactors. *Water Res. 42*, 2184-2194.

Lenz, M., & Lens, P.N.L. (2009). The essential toxin: The changing perception of selenium in environmental sciences. *Sci. Total. Environ. 407*, 3620-3633.

Levander, O.A. & Burk, R.F. (2006). Update of human dietary standards for selenium. In: Hatfield, D.L., Berry, M.J. & Gladyshev, V.N. *Selenium - its molecular biology and role in human health*. Springer, New York, USA.

Li, H., Zhang, J., Wang, T., Luo, W., Zhou, Q. & Jiang, G. (2008). Elemental selenium particles at nano-size (Nano-Se) are more toxic to Medaka (*Oryzias latipes*) as a consequence of hyper-accumulation of selenium: a comparison with sodium selenite. *Aquat. Toxicol. 89,* 251-256.

Li, D-B., Cheng, Y-Y., Wu, C., Li, W-W., Li, N., Yang, Z-C., Tong, Z-H., & Yu, H-Q. (2014). Selenite reduction by *Shewanella oneidensis* MR-1 is mediated by fumarate reductase in periplasm. *Sci. Rep. 4*, 3735.

Lovely, D. (1993). Dissimilatory metal reduction. *Ann. Rev. Microbiol. 47*, 263-290.

Lovley, D.R. (2008). Extracellular electron transfer: wires, capacitors, iron lungs, and more. *Geobiology 6*(3), 225-231.

Luoma, S.N., Johns, C., Fisher, N.S., Steinberg, N.A., Oremland, R.S., & Reinfelder, J.R. (1992). Determination of selenium bioavailability to a bivalve from particulate and solute pathways. *Environ. Sci. Technol. 26*, 485-491.

Luoma, S.N. & Presser, T.S. (2009). Emerging opportunities in management of selenium contamination. *Environ. Sci. Technol. 43*, 8483-8487.

Manceau, A. & Gallup, D.L. (1997). Removal of selenocyanate in water by precipitation: characterization of copper-selenium precipitate by x-ray diffraction, infrared, and x-ray absorption spectroscopy. *Environ. Sci. Technol. 31*, 968-976.

Maniatis, T. & Gallup, D.L. (2003). Biological Treatment of Surface and Ground Water for Selenium and Nitrate.Presented at the 2003 National Meeting of the American Society of Mining and Reclamation.Available from: http://www.asmr.us/Publications/Conference%20Proceedings/2003/0749-Maniatis.pdf.

Meng, X., Bang, S. & Korfiatis, G.P. (2002). Removal of selenocyanate from water using elemental iron. *Water Res. 36*, 3867-3873.

Merrill, D.T., Manzione, M., Parker, D., Peterson, J., Crow, W. & Hobbs, A. (1986). Field evaluation of As and Se removal by iron coprecipitation. *J. Water Poll. Control Federation 6*, 82-89.

Mollah, M.Y.A., Morkovsky, P., Gomes, J.A.G., Kesmez, M., Parga, J., & Cocke, D.L. (2004). Fundamental, present and future perspectives of electrocoagulation. *J. Hazard. Mater. 114*, 199-210.

Montgomery Watson (1995). Selenium removal from refinery wastewaters: Biological Field Testing Report. Prepared for Western States Petroleum Association, Concord, California, USA.

MSE Technology Applications, Inc. (MSE) (2001). Final report - Selenium treatment/removal alternatives demonstration project. Mine Waste Technology Program Activity III, Project 20. Report prepared for US Environmental Protection Agency, National Energy Technology Laboratory, Office of Research and Development, Cincinnati, OH and US Department of Energy, Federal Energy Technology Center, Pittsburgh, PA, USA. Available from: http://nepis.epa.gov/Adobe/PDF/P1008GVL.pdf.

Nagpal, N.K. & Howell, K. (2001). Water quality guidelines for selenium.Water protection branch.Water, Lands and Air Protection.Ministry of the Environment, Environmental Protection Division.
Available from:
http://www.env.gov.bc.ca/wat/wq/BCguidelines/selenium/selenium.html.

Nelson, B.N., Cellan, R., Mudder, J., Whitlock, J. & Waterland, R. (2003). In situ, anaerobic, biological immobilization of uranium, molybdenum and selenium in an alluvial aquifer. *Mining Eng. 55*, 31-36.

Nerenberg, R. & Rittmann, B.E. (2004). Reduction of oxidized water contaminants with a hydrogen-based, hollow-fiber membrane biofilm reactor. *Water Sci. Technol. 49*, 223-230.

Nishimura, T. & Hashimoto, H. (2007). Removal of selenium (VI) from aqueous solution with polyamine-type weakly basic ion exchange resin. *Sep. Sci. Technol. 42*, 3155-3167.

Nitrogen and Selenium Management Program (NSMP) (2007). Identification and assessment of selenium and nitrogen treatment technologies and best management practices. Orange County, CA. Available from:
http://www.ocnsmp.com/team/NSMP%20Task%202.2%20Update%2030Mar07.pdf.

North American Metal Council (NAMC) (2010). Review of available technologies for removal of selenium from water. Available from: http://www.namc.org/docs/00062756.PDF

Sura-de Jong, M., Reynolds, J., Richterova, K., Musilova, L., Staicu, L.C., Hrochova, I., Cappa, J.J., van der Lelie, D., Frantik, T., Sakmaryova, I., Strejcek, M., Cochran, A., Lovecka, P. & Pilon-Smits, E.A.H. (2015). Selenium hyperaccumulators harbor a diverse endophytic bacterial community characterized by high selenium tolerance and growth promoting properties. *Front. Plant Sci.* 6, 113.

Ohlendorf, H.M. (1989). Bioaccumulation and effects of selenium in wildlife. In: *Selenium in Agriculture and the Environment*. Jacobs, L.M. (Ed.). *Soil Science Society of America Spcl Publ* No. 23, 133-177.

Opara, A., Peoples, M.J., Adams, J.D., & Martin, A.S. (2014). Electro-biochemical reactor (EBR) technology for selenium removal from British Columbia's coal-mining wastewaters. Available from: http://www.inotec.us/uploads/5/1/2/8/5128573/selenium_removal_coal_mine_wat er_inotec-sme2014.pdf

Oremland, R.S., Hollibaugh, J.T., Maest, A.S., Presser, T.S., Miller, L.G. & Culbertson, C.W. (1989). Selenate reduction to elemental selenium by anaerobic bacteria in sediments and culture: biogeochemical significance of a novel, sulfate independent respiration. *Appl. Environ. Microbiol. 55*, 2333-2343.

Oremland, R.S. (1993). Biogeochemical transformations of selenium in anoxic environments. In: *Selenium in the environment*. Frankenberger Jr., W.T. & Dekker, M. CRC Press, New York, USA.

Pickett, T., Sonstegard, J. & Bonkoski, B. (2008). Using biology to treat selenium. *Power Eng. 110*, 140-145.

Pilon-Smits, E.A.H. & Quinn, C.F. (2010). Selenium metabolism in plants. In: *Cell biology of metal and nutrients*. Hell, R. & Mendel, R.R., eds. p. 225-241, Springer, New York, USA.

Plant J.A., Kinniburgh, D.G., Smedley, P.L., Fordyce, F.M., & Klinck, B. (2003). Treatise on Geochemistry.Vol. 9. *Arsenic and Selenium*, Editor: Sherwood Lollar, B. Elsevier, Amsterdam, The Netherlands.

Presser, T.S. & Luoma, S.N. (2006). Forecasting selenium discharges to the San Francisco Bay-Delta Estuary: Ecological effects of a proposed San Luis Drain extension. U.S. Geological Survey Professional Paper 1646.

Purkerson, D.G., Doblin, M.A., Bollens, S.M., Luoma, S.N., & Cutter, G.A. (2003). Selenium in San Francisco Bay zooplankton: Potential effects of hydrodynamics and food web interactions. *Estuaries 26*, 956-969.

Richens, D.T. (1997). *The chemistry of aqua ions*. Wiley, Chichester, UK.

Salton Sea Restoration Program (2005). Final Technologies and Management Techniques to Limit Exposure to Selenium. Available from: http://www.water.ca.gov/saltonsea/historicalcalendar/wg/03.04.2005/SeleniumMg mtTech.pdf

Schröder, I., Rech, S., Krafft, T. & Macy, J.M. (1997). Purification and characterization of the selenate reductase from *Thauera selenatis. J. Biol. Chem. 272*(38), 23765-23768.

Seby, F. (1999). Stockage des déchets nucléaires en profondeur - Etude des procédés d'immobilisation du sélénium et de l'étain, ANDRA report C RP 0LCB 99-001, 118 p.

Shamas, J., Wagner, C. & Cooke, T. (2009). Technologies and strategies for the treatment of selenium as a microconstituent in industrial wastewater.*WEF Microconstituents and Industrial water quality specialty conference*, Baltimore, MD, USA.

Schlekat, C.E., Dowdle, P.R., Lee, B.G., Luoma, S.N., & Oremland, R.S. (2000). Bioavailability of particle-associated selenium on the bivalve *Potamocorbila amuresis*. *Environ. Sci. Technol. 34*, 4504-4510.

Simmons, D.B., & Wallschlaeger, D. (2005). A critical review of the biogeochemistry and ecotoxicology of selenium in lotic and lentic environments. *Environ. Toxicol. Chem. 24*, 1331-1343.

Sobolewski, A. (2005). Evaluation of Treatment Options to Reduce Water-Borne Selenium at Coal Mines in West-Central Alberta. Report prepared for Alberta Environment Water Research Users Group Edmonton. Pub – no.T/860.

Sobolewski, A. (2013). Evaluation of Treatment Options to Reduce Water-Borne Selenium at Coal Mines in West-Central Alberta.Microbial Technologies, Inc.
Available from: http://environment.gov.ab.ca/info/library/7766.pdf.

Soda, S., Kashiwa, M., Kagami, T., Muroda, M., Yamashita, M. & Yke, M. (2011). Laboratory-scale bioreactors for soluble selenium removal from selenium refinery wastewater using anaerobic sludge. *Desalination 297*, 433–438.

Staicu, L.C., van Hullebusch, E.D., Lens PNL, Pilon-Smits, E.A.H. & Oturan, M.A. (2015a). Electrocoagulation of colloidal biogenic selenium. *Environ. Sci. Pollut. Res. Int.* 22(4), 3127-3137.

Staicu, L.C., van Hullebusch, E.D., Oturan, M.A., Ackerson, C.J., & Lens, P.N.L. (2015b). Removal of colloidal biogenic selenium from wastewater. *Chemosphere* 125, 130-138.

Staicu, L.C., Ackerson, C.J., Cornelis, P., Ye, L., Berendsen, R.L., Hunter, W.J., Noblitt, S.D., Henry, C.S., Cappa, J.J., Montenieri, R.L., Wong, A.O., Musilova, L., Sura-de Jong, M., van Hullebusch, E.D., Lens, P.N.L., Pilon-Smits, E.A.H. (2015c). *Pseudomonas moraviensis* subsp. stanleyae: a bacterial endophyte capable of efficient selenite reduction to elemental selenium under aerobic conditions (*J. Appl. Microb.*, in revision).

Stolz, J.F. & Oremland, R.S. (1999). Bacterial respiration of arsenic and selenium. *FEMS Microbiol. Rev. 23*(5), 615-627.

Tennessee Valley Authority (TVA) (2009). Environmental Assessment: initial emergency response actions for the Kingston fossil plant ash dike failure, Roane County, Tennessee.Available from:
http://www.tva.gov/environment/reports/Kingston/pdf/2009-
13_KIF_EmergencyResponse_EA.pdf

Twidwell, L., McClosky, J., Miranda, P. & Gale, M. (2000).Technologies and potential technologies for removing selenium from process and mine wastewater. Proceedings Minor Elements 2000, SME, Salt Lake City, UT, USA, pp 53-66.

United States Bureau of Reclamation (USBR) (2002). Imperial Valley: Drainwater Reclamation and Reuse Study. Available from:
http://www.usbr.gov/lc/region/g2000/publications/Imperial.pdf.

United States Bureau of Reclamation (USBR) (2008).Selenium treatment of irrigation drainage water- Options and Limitations. Available from: http://lib.berkeley.edu/WRCA/WRC/pdfs/SD08Irvine.pdf.

United States Environmental Protection Agency (USEPA) (2009). Steam Electric Power Generating Point Source Category: Final Detailed Study Report. EPA-821-R-09-008. Washington D.C., USA.

United States Environmental Protection Agency (USEPA) (2010). North San Francisco Bay selenium characterization study plan (2010-2012). Available from: http://www2.epa.gov/sites/production/files/documents/epa-r09-ow-2010-0976-0023-1.pdf.

United States Environmental Protection Agency (USEPA) (2013). Basic Information about Selenium in Drinking Water. Available from: http://water.epa.gov/drink/contaminants/basicinformation/selenium.cfm.

Van Ginkel, S.W., Yang, Z., Kim, B.O., Sholin, M., & Rittmann, B.E. (2011a). Effect of pH on nitrate and selenate reduction in flue gas desulfurization brine using the H_2-based membrane biofilm reactor (MBfR). *Water. Sci. Technol. 63 (*12), 2923-2928.

Van Ginkel, S.W., Yang, Z., Kim, B.O., Sholin, M., & Rittmann, B.E. (2011b). The removal of selenate to low ppb leves from flue gas desulfurization brine using the H_2-based membrane biofilm reactor (MBfR). *Bioresour. Technol. 102* (10), 6360-6364.

Watts, C.A., Ridley, H., Condie, K.L., Leaver, J.T., Richardson, D.J. & Butler, C.S. (2003). Selenate reduction by *Enterobacter cloacae* SLD1a-1 is catalysed by a molybdenum-dependent membrane-bound enzyme that is distinct from the membrane-bound nitrate reductase. *FEMS Microbiol. Lett. 228*(2), 273-279.

Wen, H. & Carignan, J. (2007). Reviews on atmospheric selenium: emissions, speciation and fate. *Atmos. Environ. 41*, 7151-7165.

World Health Organisation (WHO) (2011). Guidelines for Drinking-water Quality. 4th Edition. Available from: http://whqlibdoc.who.int/publications/2011/9789241548151_eng.pdf.

Ye, Z.H., Lin, Z.Q., Whiting, S.N., de Souza, M.P. & Terry, N. (2003). Possible use of constructed wetland to remove selenocyanate, arsenic, and boron from electric utility wastewater. *Chemosphere 52*, 1571-1579.

Yudovich, Ya.E. & Ketric, M.P. (2005). Selenium in coal: A review. *Int. J. Coal Geol. 67*, 112-126.

Zhang, Y., Zahir, Z.A., & Frankenberger Jr., W.T. (2004). Fate of colloidal-particulate elemental selenium in aquatic systems. *J. Environ. Qual. 33*, 559-564.

Zhang, Y., Wang, J., Amrhein, C. & Frankerberger Jr., W.T. (2005). Removal of selenate from water by zerovalent iron. *J. Environ. Qual. 34*, 487-495.

Zhang, E.S., Wang, H.L., Yan, X.X. & Zhang, L.D. (2005). Comparison of short-term toxicity between Nano-Se and selenite in mice. *Life Sc.i 76*, 1099-1110.

CHAPTER 3

Pseudomonas moraviensis subsp. stanleyae: a bacterial endophyte capable of efficient selenite reduction to elemental selenium under aerobic conditions

This chapter was published as:

Staicu, L.C., Ackerson, C.J., Cornelis, et al. (2015). *Pseudomonas moraviensis* subsp. stanleyae: a bacterial endophyte capable of efficient selenite reduction to elemental selenium under aerobic conditions. Journal of Applied Microbiology (accepted).

Chapter 3. *Pseudomonas moraviensis* subsp. stanleyae: a bacterial endophyte capable of efficient selenite reduction to elemental selenium under aerobic conditions

Abstract

Aims: Identification of bacteria with high selenium tolerance and reduction capacity, for bioremediation of wastewaters and nanoselenium particle production.

Methods and Results: A bacterial endophyte was isolated from the selenium hyperaccumulator *Stanleya pinnata* (Brassicaceae) growing on seleniferous soils in Colorado, USA. Based on Fatty Acid Methyl Ester (FAME) analysis and Multi-locus Sequence Analysis (MLSA) using *16S rRNA, gyrB, rpoB* and *rpoD* genes, the isolate was identified as a subspecies of *Pseudomonas moraviensis* (97.3% nucleotide identity) and named *P. moraviensis* stanleyae. The isolate exhibited an extremely high tolerance to SeO_3^{2-} (up to 120 mM) and SeO_4^{2-} (>150 mM). Selenium oxyanionremoval from growth medium was measured by Microchip Capillary Electrophoresis (detection limit 95 nM for SeO_3^{2-} and 13 nM for SeO_4^{2-}). Within 48 h, *P. moraviensis* stanleyae aerobically reduced SeO_3^{2-} to red Se(0) from 10 mM to below the detection limit (removal rate 0.27 mM h^{-1} at 30 °C); anaerobic SeO_3^{2-}removal was slower. No SeO_4^{2-}removal was observed. *P. moraviensis* stanleyae stimulated growth of crop species *Brassica juncea* by 70% with no significant effect on Se accumulation.

Conclusions: *P. moraviensis* stanleyae can tolerate extreme levels of selenate and selenite and can deplete high levels of selenite under aerobic and anaerobic conditions.

Significance and Impact of the study: *Pseudomonas moravensis* subsp. stanleyae showing high efficiency in aerobic reduction of selenite may be proposed for the abatement of this toxic Se oxyanion in wastewaters and for the growth enhancement of several crop species.

Keywords: Aerobic selenite reduction; *Pseudomonas moraviensis*; Elemental selenium nanoparticles; *Stanleya pinnata*; MLSA; MCE.

3.1. Introduction

Selenium (Se) is an element found in fossil fuels, phosphate deposits, sulfide minerals and seleniferous soils (Lemly, 2004). Its complex biochemistry allows Se to be cycled through different environmental compartments (Chapman et al., 2010). The two oxyanions of Se, selenate (Se[VI], SeO_4^{2-}) and selenite (Se[IV], SeO_3^{2-}), are water-soluble, bioavailable and toxic (Simmons & Wallschlaeger, 2005). Because of the toxicity posed by Se oxyanions, the U.S. Environmental Protection Agency has set a limit of 50 µg L^{-1} for Se in drinking water (USEPA, 2003). In its elemental state, Se(0), Se is water-insoluble and less bioavailable

(Chapman et al., 2010). Various industrial sectors produce wastewaters containing toxic Se oxyanions that can be cleaned up using a microbial treatment system, provided the bacterial inoculum can reduce Se oxyanions to solid Se(0), that can be further removed from wastewater (Sobolewski, 2013; Staicu et al., 2015). Considering the future trends in energy production based on fossil fuel combustion, it is expected that Se will increase its presence and toxicity in the environment (Lenz & Lens, 2009). To cope with this challenge, biotechnological Se removal processes can be employed as a cheaper and more efficient alternative over physical-chemical clean-up technologies (NAMC, 2010).

Several *Pseudomonas* species have been reported to metabolize Se oxyanions (*P. seleniipraecipitatus* in Hunter & Manter (2011); *P. stutzeri* NT-I in Kuroda et al. (2011) but with different degrees of success. The genus *Pseudomonas* encompasses a wide array of genetically and metabolically diverse bacterial species. Since Walter Migula coined and introduced the name *Pseudomonas* in 1894, the genus has undergone a dramatic rearrangement to the 213 species indexed at the time of writing (http://www.bacterio.cict.fr/p/pseudomonas.html). Different *Pseudomonas* species have been shown to be opportunistic human pathogens of clinical relevance (*P. aeruginosa*), plant pathogens (*P. syringae*), denitrifiers (*P. stutzeri*), or plant growth promoters (*P. fluorescens, P. putida, P. chlororaphis*) (Spiers et al., 2000; Zago & Chugani, 2009). *Pseudomonas* representatives are rod-shaped, Gram negative γ Proteobacteria, motile by one or more polar flagella, aerobic, catalase positive, and chemoorganotrophic, but all these characteristics do not allow an absolute differentiation (Palleroni, 2005; Peix et al., 2009). The determination of sequence similarity in ribosomal RNA is accepted as a solid classification argument. However, in the case of the genus *Pseudomonas*, the discriminative power of 16S sequencing is rather low, and now a multi-locus sequencing analysis approach combining different housekeeping genes is preferred (Mulet et al., 2010).

In the current work, we describe the isolation and characterization of a bacterial endophyte (*Pseudomonas moraviensis* stanleyae) that dwells in the roots of *Stanleya pinnata,* a selenium hyperaccumulator plant species. Hyperaccumulators are plants that accumulate one or more chemical elements to levels typically two orders of magnitude higher than those in the surrounding vegetation (Terry & Banuelos, 2000). *Stanleya pinnata* is native to Se-rich soils throughout the Western USA and has been shown to reach up to 2,000 mg kg^{-1} Se (0.2%) in its root tissues, and even higher levels in leaves and flowers (Freeman et al., 2006; Galeas et al., 2007). The bacterial strain, derived from this extreme environment and reported herein, was tested for selenite and selenate tolerance and reduction under aerobic and anaerobic conditions.

3.2. Materials and methods

3.2.1. Media and culture conditions

Luria Broth (LB) was purchased from Fisher Scientific. When the growth media was amended with salts of Se oxyanions, the aliquots were added from filter-sterilized stock solutions (1 mol L^{-1}). Sodium selenate, Na_2SeO_4, ≥ 98%, and sodium selenite, Na_2SeO_3, ≥ 99%, were purchased from Sigma Aldrich. All other reagents were of analytical grade unless otherwise stated. The incubations were performed at 30 °C, pH 7.5, 200 rpm, under aerobic conditions. For anaerobic incubations, rubber septa serum bottles containing LB media were used and the headspace of the sealed bottles was replaced and purged with N_2 for 15 min prior to incubation at 30 °C, pH 7.5, and 200 rpm.

3.2.2. Isolation of strain #71

The strain, with accession #71, was isolated from the root tissue of Se hyperaccumulator *Stanleya pinnata* (*Brassicaceae*), growing on the seleniferous soils of Pine Ridge Natural Area (Colorado, USA) on the west side of Fort Collins (40°32.70 N, 105°07.87 W, elevation 1,510 m). Plants were harvested in the field during August 2012 and brought to the lab for storage at 4 °C before further processing.

After washing, the plants were surface-sterilized for 15 min on a rotary shaker using 2% sodium hypochlorite (NaClO) and 0.5 mL L^{-1} Tween 20. This step was followed by three washings with sterile H_2O. The last rinsing water was plated on half-strength LB media as a control measure, to test for surface sterility. The tissues were transferred to 10 mL of sterile 10 mM $MgSO_4$ andthen ground at room temperature under sterile conditions using a micropestle and microcentrifuge tubes. The homogenate was allowed to settle and separate gravitationally and the supernatant sampled. 100 µL aliquots of the supernatant were placed on solid half-strength LB media. The incubation was performed at room temperature for seven days. Individual colonies were sub-cultured on new media to gain pure bacterial monocultures. To test for SeO_3^{2-} reduction capacity, bacterial monocultures were sub-cultured at 30 °C on the same solid media containing 10 mM sodium selenite. Amongst monocultures, isolate #71 was selected for this study based on its apparent high Se tolerance and ability to reduce selenite to red elemental Se.

3.2.3. Identification of strain #71

Whole-cell fatty acid methyl ester (FAME) analysis

For Fatty Acid Methyl Esters analysis, 48 h pure cultures cultivated on slant LB agar were supplied to MIDI Labs Inc. (Newark, DE). The results were analyzed using the Sherlock Microbial Identification System 6.2.

Multi-locus sequence analysis

The taxonomic position of the isolate was investigated using a Multi-locus sequence analysis (MLSA), as described by (Mulet et al., 2010). The concatenated partial sequences of four housekeeping genes (*16S rRNA, gyrB, rpoB* and *rpoD*) of the strain (Table 3.1) were aligned with those of 107 *Pseudomonas* type strains. A phylogenetic tree was generated based on the alignment by neighbor-joining using the CLC main workbench 6.7.2 (CLC bio, Aarhus, Denmark). 16S rRNA's were amplified using universal primers 16F27 and 16R1492 (Table 3.1). Housekeeping genes were amplified using the following primers: UP1E, APrU, M13R and M13(-21) (for *gyrB*); LAPS5 and LAPS27 (for *rpoB*); and PsEG30F and PsEG790R (for *rpoD*) (Table 3.1). The PCR conditions were as follows: 94 °C for 5 min; then 94 °C for 1 min, 57 °C for 45 s, 72 °C for 2 min for 30 cycles; then 72 °C for 10 min for final extension followed by cool down to 4 °C. The DNA fragments were analyzed in 1% agarose gel.

Table 3.1. Multi-locus sequence analysis parameters.

Gene	Primer	Sequence (5'- 3')	Length (bp)	Reference	Fragment length (bp)
16S rRNA	16F27	AGAGTTTGATCMTGGCTCAG	20	Lane, 1991	1465
	16R1492	TACGGYTACCTTGTTACGACTT	22	Lane, 1991	
gyrB	UP1E	CAGGAAACAGCTATGACCAYGSNGGNGGNA RTTYRA	36	Yamamoto et al., 2000	966
	APrU	TGTAAACGACGGCCAGTGCNGGRTCYTTYTCY TGRCA	37	Yamamoto et al., 2000	
	M13R	CAGGAAACAGCTATGACC	18	Yamamoto et al., 2000	
	M13(-21)	TGTAAACGACGGCCAGT	17	Yamamoto et al., 2000	
rpoB	LAPS5	TGGCCGAGAACCAGTTCCGCGT	22	Tayeb et al., 2005	1229
	LAPS27	CGGCTTCGTCCAGCTTGTTCAG	22	Tayeb et al., 2005	
rpoD	PsEG30F	ATYGAAATCGCCAARCG	17	Mulet et al., 2009	760
	PsEG79R	CGGTTGATKTCCTTGA	16	Mulet et al., 2009	

3.2.4. Growth test

Isolate #71 was grown aerobically in liquid LB media at 30 °C under constant shaking (200 rpm) and the samples were analyzed every 3 hours during the first day. The growth curve was constructed based on plate colony count (Colony Forming Units, CFU) and optical density (OD) measurements at 600 nm absorbance using a Beckman DU530 spectrophotometer (Table 3.2). When grown in the presence of 10 mM Na_2SeO_3, to avoid the spectral interference of red elemental Se, an indirect method was employed. Red Se(0) absorbs at a wavelength around 612 nm (Kumar et al., 2014), therefore the production of Se(0) particles interferes with the spectrophotometric (OD_{600}) bacterial growth measurements. Following a procedure adapted from (Di Gregorio et al., 2005) a control experiment was used to measure the OD_{600} and 200 μL of a 10^{-6} dilution of each time point (6, 9, 12, 24 and 48 h) was plated on solid LB media and incubated at 30 °C for 24 h. The experiment containing Na_2SeO_3 was also 10^{-6} diluted and plated together with the diluted control under the same conditions. After 24 h, the CFU number that developed on the control and selenite-amended agar plates was counted and the results (Table 3.2) were statistically processed (mean and standard deviation) using SigmaPlot. The growth curve of the selenite-containing treatment (in triplicate) was built by back calculation using the correlated OD_{600}-CFU of the control treatment.

Table 3.2. Growth of *P. moraviensis* stanleyae based on Colony Forming Units (CFU) and absorbance at OD_{600} in the absence and presence of 10 mM of Na_2SeO_3 and Na_2SeO_4.

Time (h)	Control (OD_{600})	Control (CFU)	SD	10^8 CFU / mL	SeO_3^{2-} (CFU)	SD	10^8 CFU / mL	SeO_4^{2-} (OD_{600})
0	0	ND	ND	ND	ND	ND	ND	0
3	0.1645	ND	ND	ND	ND	ND	ND	0.207
6	0.668	78	10	3.9	18	5	0.9	0.689
9	1.060	102	13	5.1	55	8	2.8	1.012
12	1.266	203	21	10	135	18	6.8	1.181
24	1.343	282	12	14	133	8	6.7	1.292
48	1.396	296	16	15	134	11	6.7	1.433

3.2.5. Selenium oxyanions measurement

Inorganic selenite concentrations were measured using microchip capillary electrophoresis (MCE) in a poly(dimethylsiloxane) (PDMS) device with contact conductivity detection (Noblitt & Henry, 2008). The separation background electrolyte was recently developed to specifically target Se oxyanions with high selectivity and sensitivity (Noblitt et al., 2014). Detection limits are about 53 and 280 nM for selenate and selenite, respectively. Samples were diluted 250-fold in background electrolyte prior to analysis, yielding respective detection limits of 13 µM and 95 µM. The specific selenite reduction rate was calculated from the slope of the linearized time course.

3.2.6. Selenium tolerance

Na_2SeO_3 and Na_2SeO_4 were added to LB media at increasing concentrations from 0.1 to 150 mM. The first concentrations used were 0.1, 0.5, 1, 5 and 10 mM. Between 10 and 150 mM, the Se oxyanions concentrations were increased by 10 mM-increments. Each test tube was inoculated with 1% (v/v) of the same stock culture of *P. moraviensis* stanleyae. The Se tolerance was determined by the highest SeO_3^{2-} and SeO_4^{2-} concentration at which growth was detected by spectrophotometric measurement at OD_{600}. In the case of SeO_3^{2-}, growth was accompanied by the production of red Se(0).

3.2.7. Transmission Electron Microscopy (TEM)

For TEM, 24-hour cultures grown in LB containing either 10 mM Na_2SeO_3 or no Na_2SeO_3 (control) were sampled, processed and fixed in a solution containing 2.5% glutaraldehyde and 2% formaldehyde (Mishra et al., 2011). Aliquots of 5 µL were pipetted onto 400 mesh carbon coated copper TEM grids (EM Sciences). Excess liquid was wicked off with filter paper after one min. The resulting samples were examined in a JEOL JEM-1400 TEM operated at 100 kV and spot size 1. The Se(0) particle size was determined by TEM image processing using ImageJ[TM] 1.47v software.

3.2.8. Inoculation experiment

Indian mustard (*Brassica juncea* L.) plants were grown from surface-sterilized seeds on soil collected from Pine Ridge Natural Area (for soil properties see Galeas et al., 2007). The soil was collected in the field and mixed with Turface® gravel in a 2:1 soil:Turface® ratio. Polypropylene (Magenta) boxes were filled to a height of 2 cm with this mixture, and autoclaved for 40 min. Seeds were surface-sterilized by rinsing for 30 min in 15% household bleach (1.5% NaClO) followed by five 5 min rinses in sterile water, and then sown in the Magenta boxes at a density of 3 seeds per box and six boxes per treatment. One week after germination, the seedlings were thinned to one plant per box and inoculated with *P. moraviensis* stanleyae (#71); there was an uninoculated parallel control treatment. Before inoculation, the bacteria were grown in half-strength LB for 24h at 25 °C, harvested by centrifugation and resuspended in 10 mM $MgSO_4$ to an OD_{600} of 1.0. One mL of inoculum was delivered using a pipette to the base of each seedling; the controls received 1 mL of 10 mM $MgSO_4$. The plants were allowed to grow for 6 weeks. The boxes were watered with autoclaved water every 2 weeks (twice total) in a laminar flow sterile hood. The plants were then harvested, separating the root and shoot. Small shoot and root samples from each plant were placed in 10 mM $MgSO_4$ for re-isolation of bacterial endophytes, to verify successful inoculation. These were ground using sterile micropestles in microcentrifuge tubes, and 100 µL of the extract was streaked onto LB agar plates, which were monitored after 24h and compared visually with the inoculum. The remainder of the root and shoot material was dried and weighed. Root and shoot samples were digested in nitric acid according to Zarcinas et al. (1987) and analyzed for elemental composition using inductively coupled plasma optical emission spectrometry (ICP-OES) according to Fassel (1978).

3.2.9. Statistical analysis

The results were statistically processed and plotted using the data analysis software SigmaPlot 12.0v. When the standard deviation was less than 5%, the error bars are not presented in the figures. All experiments were performed in triplicate unless otherwise stated.

3.3. Results

3.3.1. Phylogenetic analysis

After isolation of endophytic bacteria from surface-sterilized *S. pinnata* roots, strain #71 was selected for further study based on its apparent tolerance to selenite and production of red elemental Se. In order to identify the phylogenetic position of strain #71, we performed a multiphase analysis. First, the cellular fatty acid content of the strain was analyzed (Table 3.3). The major fatty acids were $C_{16:1}$ w7c/16:1 w6c, $C_{16:0}$, and $C_{18:1}$w7c.

Table 3.3. Fatty Acid Methyl Esters (FAME) profile of *Pseudomonas moraviensis* stanleyae.

Fatty acid	(%)
3-OH $C_{10:0}$	3.42
$C_{10:0}$	1.61
2-OH $C_{12:0}$	6.18
3-OH $C_{12:0}$	5.03
$C_{14:0}$	0.55
$C_{16:0}$	29.85
$C_{16:1}$ w7c/16:1 w6c	35.87
$C_{17:0}$ cyclo	3.00
$C_{18:0}$	0.42
$C_{18:0}$ ante/18:2 w6, 9c	0.23
$C_{18:1}$ w7c	13.84

To further establishthe taxonomic position of strain #71, we amplified four housekeeping genes (16S rRNA, *gyrB*, *rpoB* and *rpoD*)and compared the concatenated sequences of these genes to those of 107 *Pseudomonas* type strains as described by (Mulet et al., 2010). Figure 3.1 shows the phylogenetic position of strain #71.

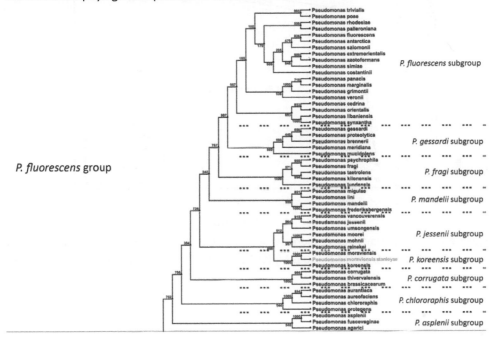

Figure 3.1. Neighbor-joining tree based on the 16S rRNA, *gyrB*, *rpoB* and *rpoD* gene sequences of *P. moraviensis* stanleyae and 107 *Pseudomonas* type strains. The phylogenetic position of *P. moraviensis* stanleyae within the *P. fluorescens* group is shown in blue.

3.3.2. Growth

The aerobic growth of *P. moraviensis* stanleyae in LB media with and without the Se-oxyanions selenite and selenate is presented in Figure 3.2. The control (no Se) exhibited a lag phase of around 4 h, followed by a steep logarithmic growth phase. After 9 h of growth, the culture entered the stationary phase. A similar growth curve was observed with 10 mM sodium selenate-amended media. In contrast, when 10 mM sodium selenite was present, the lag phase was prolonged to around 6 h, followed by a 6-h long exponential growth phase. The stationary phase of the SeO_3^{2-}-treatment started after around 12 h of incubation.

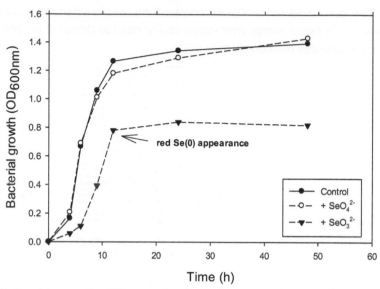

Figure 3.2. Aerobic growth of *P. moraviensis* stanleyae. Aerobic growth: control (LB media without Se oxyanions); $[SeO_3^{2-}]$ = 10 mM; $[SeO_4^{2-}]$ = 10 mM; the arrow indicates the time of red Se(0) appearance.

3.3.3. SeO_3^{2-} reduction

The aerobic reduction of selenite was further investigated by means of microchip capillary electrophoresis (MCE). During the first 9 h of incubation, the initial 10 mM selenite did not decrease (Figure 3.3). After 9 h, SeO_3^{2-} was depleted linearly with time, at a reduction rate of 0.27 mM h^{-1} (R^2 = 0.999) at 30 °C. No selenite was detected in the supernatant after 48 h of incubation. The depletion of selenite was paralleled by a change in color of the culture from off-white to pink to deep red (results not shown). In contrast, the selenate concentration did not decrease even after 48 h of incubation (Figure 3.3) and also no red elemental Se was observed.

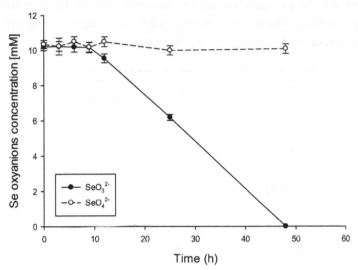

Figure 3.3. Aerobic reduction of selenium oxyanions by *P. moraviensis* stanleyae as a function of time.

3.3.4. Selenium tolerance

The *P. moraviensis* strain described in the current study showed extremely high tolerance to Se oxyanions. It grew and produced elemental Se in medium containing up to 120 mM selenite; however, above 30 mM concentration there was an increasing lag time. During the first 48 h, the strain showed growth and formation of red Se(0) in selenite concentrations up to 30 mM. During day 3, bacterial growth was detected in incubations containing up to 50 mM SeO_3^{2-}. From day 4 to day 10, bacterial growth was detected in incubations containing progressively higher selenite concentrations (from 60 mM for day 4 to 120 mM at day 10, respectively). At SeO_3^{2-} concentrations higher than 120 mM, no bacterial growth was recorded at all, until day 15. In the case of SeO_4^{2-}, bacterial growth was recorded during the tested period up to the highest concentration used, 150 mM.

3.3.5. Production of red elemental selenium

The production of red Se(0) was strongly correlated with bacterial growth (Table 3.4). Red Se(0) was produced both under aerobic and anaerobic conditions by the reduction of selenite (Fig. 3.4A), likely as a detoxification reaction (Kessi et al., 1999). Although the strain could grow in the presence of 10 mM SeO_3^{2-} or in the presence of 5% (w/v) NaCl, when both were present no growth and no red Se(0) formation were detected. At 41 °C the strain could only produce limited amounts of red Se(0), indicative of suboptimal growth conditions induced by higher temperatures.

Table 3.4. Growth of *P. moraviensis* stanleyae and formation of red Se(0) in LB media under different growth conditions. Notes: -, no growth or red Se(0) production; +, growth and red Se(0) production; +++, optimal growth and red Se(0) production.

Condition	Growth	Red Se(0) formation
Aerobic		
4 °C	-	-
28 °C	+++	+++
41 °C	+	+
NaCl 5%	+	+
NaCl 7%	-	-
Anaerobic		
28 °C	+	+
*Heat-killed inoculum**	-	-

*autoclavedat 121 °C for 15 min

Figure 3.4B presents a TEM micrograph of *P. moraviensis* stanleyae grown aerobically for 24 h in medium amended with 10 mM of Na₂SeO₃. Elemental Se nanoparticles are visible on the surface of bacterial cells and display a size lower than 100 nm.

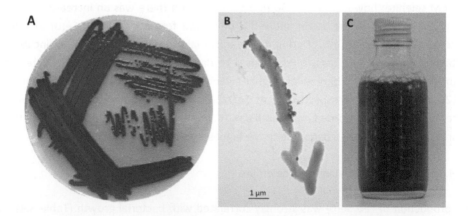

Figure 3.4. Biogenic Se(0). (A) Red elemental Se formation on agar plate (aerobic conditions), (B) Transmission Electron Micrograph (TEM) of *P. moraviensis* stanleyae producing Se(0) under aerobic conditions (the arrows indicate the presence of electron-dense Se(0) particles), and (C) Red Se(0) produced under anaerobic conditions by the reduction of [SeO₃²⁻] = 10 mM (described in Materials and Methods). The red color starts to become apparent after 24 h of incubation.

3.3.6. Inoculation

Since this *Pseudomonas* strain was a root endophyte isolated from a Se hyperaccumulator from the Brassicaceae family (*S. pinnata*), its capacity to affect growth and accumulation of Se and other plant nutrient was investigated in related crop species Indian mustard (*B. juncea*). Inoculation with *P. moraviensis* stanleyae enhanced root growth by 38% (NS) and shoot growth by 70% (P<0.05, Table 3.3). Root and shoot Se levels were not significantly affected in the inoculated plants, but showed a trend to be lower compared to the control plants (Table 3.5). In contrast, the chemically similar element sulfur (S) was present at higher levels in roots of inoculated plants compared to control plants (Table 3.3), while shoot S levels were unaffected. The effect of inoculation on plant potassium (K) levels was very similar to those of S: K levels were higher in root and unaffected in shoots of inoculated plants, as compared to control plants (Table 3.5). Calcium translocation from root to shoot appeared to be reduced in inoculated plants, since Ca levels were higher in the root and lower in the shoot of inoculated plants, compared to control plants (Table 3.5). Manganese (Mn) levels were significantly higher in both roots and shoots of inoculated plants, compared to uninoculated controls (Table 3.5). Other elements that were also tested but found to not be different between inoculated and control plants include Cu, Fe, Mg, Mo, Ni and Zn (results not shown).

Table 3.5. Dry weight (mg) and elemental concentration (mg kg^{-1} dry weight) for *Brassica juncea* grown without (control) and with (inoculated) *P. moraviensis stanleyae*. Shown values are the mean and standard error of the mean ($n = 6$). All inoculated values that are significantly different from the control (p < 0.05) are indicated by an asterisk (*).

	Root		Shoot	
	Control	Inoculated	Control	Inoculated
Dry weight	8 ± 1.4	11 ± 1.3	17 ± 3.4	29 ± 2.7*
Ca	29,221 ± 3,040	44,513 ± 15,141*	14,580 ± 2,023	9,920 ± 581*
K	14,917 ± 1,411	30,362 ± 5,669*	61,470 ± 12,331	62,193 ± 2,500
Mn	592 ± 65	1,155 ± 214*	449 ± 477	735 ± 127*
S	3,831 ± 291	7,563 ± 1,129*	17,183 ± 1997	16,396 ± 448
Se	450 ± 118	345 ± 135	1,692 ± 434	554 ± 122

3.4. Discussion

3.4.1. Phylogenetic analysis

Based on the fatty acid methyl esters of strain #71, a similarity index (SIM) of 0.846 was calculated, matching with the FAME profile of *Pseudomonas fluorescens*. According to the FAME protocol, SIM indices higher than 0.6 are indicative of a good species match.

The genus *Pseudomonas*, however, is very diverse and its taxonomy is still developing (Mulet et al., 2010). Using the multi-locus sequencing analysis, the phylogenetic position of strain #71 was further refined to *P. koreensis*, a subgroup of *P. fluorescens,* and was finally found to be most closely related to the *P. moraviensis* type strain, with 97.3% nucleotide identity (NI) match between the concatenated sequences (Figure 3.1). As 97% NI was proposed as the species boundary (Mulet et al., 2010), we classified strain #71 as a subspecies of *P. moraviensis* and named it *P. moraviensis* stanleyae in reference to its host plant. *P. moraviensis* was isolated from soil by selective enrichment with nitroaromatics (Tvrzova et al., 2006).

Taken together, these two phylogenetic approaches show the importance of the best discriminative tool towards the proper identification of a species. The genus *Pseudomonas* has been shown to be complex and often various subspecies can have contrasting metabolic traits (Palleroni, 2005). This is particularly important since different wastewaters have a complex make-up with competing oxyanions that can interfere with the target pollutants to be treated by bioremediation.

3.4.2. Growth

P. moraviensis stanleyae can grow aerobically in the presence of both selenite and selenate but exhibits different growth patterns. While the presence of selenate does not appear to have a negative impact on the growth curve compared to the control treatment, the presence of selenite elicits a toxic effect. The control and the selenate-amended treatments showed a typical sigmoid growth curve and the treatments exhibited an almost similar growth pattern. In contrast, when the culture was amended with 10 mM selenite, a 40% diminished cell concentration (based on CFU count) was measured during stationary phase. The comparison between growth phases exhibited by the selenite-amended culture and the control show an extended lag phase induced by selenite (6 h vs 4 h). Moreover, the stationary phase also starts with a 3-h delay in the selenite-amended treatment (12 h vs 9 h). This indicates that the cell division takes place at a slower rate when selenite is present.

Overall, the delay in growth phases and the slower growth of the selenite-amended culture indicate a toxic effect elicited by selenite. Similarly, a 40% lower cell concentration between the control and the 0.5 mM SeO_3^{2-}-amended phototrophic bacterium *Rhodospirillum rubrum* cultures during stationary growth phase was reported Kessi et al. (1999). It has been shown that selenite reacts with glutathione forming toxic reactive oxygen species (H_2O_2 and O_2^-) that trigger the production of further oxidative stress enzymes (Kramer & Ames, 1988; Babien et al., 2001).

It is interesting to note that, even though *Pseudomonas* is often considered an obligate aerobic genus, production of red Se(0) by *P. moraviensis* stanleyae was observed under anaerobic conditions (Fig. 3.4C). The fact that selenite was reduced anaerobically might indicate its possible use as a terminal electron acceptor when oxygen concentration is depleted, in analogy with *Bacillus selenitireducens* (Switzer Blum et al., 1998).

3.4.3. SeO_3^{2-} reduction

The poor depletion of selenite during the first 9 h of incubation can be linked to the longer time needed by the culture to reach the stationary phase. The high selenite reduction showed during the stationary phase points out the importance of keeping the inoculum in a steady stationary phase in order to ensure the efficient and high rate conversion of selenite. Furthermore, the short period needed until the stationary phase is reached suggests that the strain under investigation could be a promising alternative to the slower anaerobic inocula. Compared with the anaerobic reduction of selenite by mixed anaerobic microbial communities (granular sludge) in a batch mode, which may take 10-14 days (Lenz & Lens, 1999), *P. moraviensis* stanleyae is at least 5-fold faster under aerobic conditions. Using *Shewanella oneidensis* MR-1 grown anaerobically at 30°C, in LB media containing 0.5 mM selenite and 20 mM fumarate, Li et al. (2014) reported selenite reduction rates between 0.5 and 1.5 μM h^{-1} for the wildtype and three mutants. As a comparison, the 0.27 mM h^{-1} selenite reduction rate reported herein is 3 orders of magnitude faster. This strain therefore holds promise for the development of a more efficient aerobic treatment system of selenite-laden wastewaters.

The depletion of selenite below the highly sensitive detection limits of MCE is particularly relevant in the context of the very low (50 μg L^{-1}) permissible discharge limits imposed by the regulatory agencies (USEPA, 2003).

3.4.4. Tolerance to selenite and selenate

The amount of selenite that can be reduced to red Se(0) is an important parameter in the design and operation of bioreactors treating Se-laden wastewaters. With increasing selenite concentration, the culture exhibited a longer delay before the onset of the reduction process and the formation of red Se(0). Similar high tolerance to selenite was reported for other bacteria (Kuroda et al., 2011; Lampis et al., 2014). A *Pseudomonas stutzeri* strain collected from the drainage water of a Se refinery plant (Kuroda et al., 2011) was reported to aerobically reduce high concentrations of selenite (up to 94 mM), although the higher levels were not completely depleted and selenite could not be reduced anaerobically. In contrast, other Pseudomonas strains were reported to be Se sensitive (Ike et al., 2000). This entails that different strains are adapted to environments with different evolutionary pressures (Rajkumar et al., 2012).

In contrast, within the selenate range investigated, 0.1 to 150 mM, growth of *P. moraviensis* stanleyae was not inhibited. Since selenate did not negatively impact the growth of the strain at any of the concentrations tested, the maximum SeO_4^{2-} that it can withstand could not be determined and it can be assumed to be higher than 150 mM. Growth in the presence of SeO_4^{2-} was not accompanied by the production of red Se(0). Thus, the resistance of *P. moraviensis* stanleyae to selenate is not based on selenate reduction, but may be due to the extrusion of this oxyanion by membrane efflux systems (Bruins et al., 1999). Only a limited number of bacterial species have been shown to reduce SeO_4^{2-} to red Se(0) under aerobic conditions (reviewed in Kuroda et al., (2011)).

The high Se tolerance and selenite reduction capacity exhibited by *P. moraviensis* stanleyae may be correlated to the seleniferous soils from the Pine Ridge Natural Area (Colorado) from which it originates, and the extremely high Se levels in its host plant (El Mehdawi & Pilon-Smits, 2012). Selenium hyperaccumulator *S. pinnata* typically contains tissue Se levels upwards of 2,000 mg kg^{-1} dry weight in all its organs (Galeas et al., 2007; Cappa et al., 2014). In *S. pinnata*, Se is stored predominantly in organic forms, which it reductively assimilates from selenate (Freeman et al., 2006). From an evolutionary standpoint, bacteria living in the rhizosphere or as endophytes of various metallophytes may be expected to develop tolerance to the toxic elements that are tolerated by their host plants. Apart from their intrinsic scientific merit, these plant-microbe symbioses constitute a promising source for bacteria with favorable properties for industrial applications. In our further studies with *P. moraviensis* stanleyae we intend to explore such applications.

3.4.5. Production of red elemental selenium

The formation of red Se(0) was dependent on bacterial growth. The aerobic and anaerobic negative controls (no bacterial inoculum added) amended with selenite did not turn red, therefore the formation of red Se(0) was mediated by bacterial activity and not by an abiotic process (results not shown). Because Se(0) is water-insoluble and less bioavailable than Se oxyanions, it has reduced toxicity (White et al., 1997). This principle is used by bacteria to render Se oxyanions less harmful (Kessi et al., 1999) and may be similarly applied in industry. Extracellular Se particles of similar size ($\sim$ 120 nm) were reported on the surface of *Veillonella atypica* cells grown in the presence of 5 mM Na_2SeO_3 (Pearce et al., 2008). The particles may be formed as a result of selenite reduction in or outside the cell (Oremland et al., 2004).

Bacteria that reduce Se oxyanions may be used in bioreactors to treat Se-laden wastewaters, such as the selenite-polluted waters produced by power plants (Hansen et al., 1998). In addition, the conversion of high levels of selenite to elemental Se(0) exhibited by the strain investigated in the current study opens the possibility to gain further insight into the production of biogenic Se nanoparticles. Further research using high-resolution Scanning Electron Microscopy (SEM) is needed to assess the level of polydispersity of biogenic Se(0) nanoparticles. Having *photoconductive* properties (i.e. electrical resistance decreases with the increase in illumination) as well as *photovoltaic* properties (i.e. the direct conversion of light into electricity), elemental Se is routinely exploited in a variety of industrial applications including solar and photo cells, exposure meters, and xerography (Johnson et al., 1999).

Another interesting alternative would be the production of binary metal compounds like quantum dots (e.g. CdSe) (Pearce et al., 2008). Ayano et al. (2014) produced CdSe nanoparticles by incubating cadmium chloride and sodium selenite in the presence of a *Pseudomonas* isolate with high metal tolerance. However, the grand challenge of this approach is the need for a complete reduction of selenite (Se^{4+}) to selenide (Se^{2-}). If the reduction will stop at Se(0), the yield of CdSe will be decreased.

3.4.6. Inoculation experiment

The root endophyte *P. moraviensis* stanleyae enhanced the growth of Indian mustard, a crop relative of its natural host, *S. pinnata*. This indicates that likely *P. moraviensis* stanleyae is a mutualistic symbiont and that its host range is not limited to the Se hyperaccumulator *S. pinnata*. The finding that this bacterial species could boost plant productivity of Indian mustard by 70% is of significance since it likely can also enhance growth of other economically important Brassicaceae crop species such as canola, cabbage, broccoli, radish, turnip and others. The mechanism underlying the stimulation of plant growth by this bacterium awaits further study, but might involve the production of the growth hormone indole acetic acid. *P. moraviensis* stanleyae enhanced plant accumulation of the plant nutrients Mn, S and K, and also affected Ca distribution in the plant. Selenium, the main element of interest, was not significantly affected, although it is noteworthy that the mean shoot Se concentration was 3-fold lower in the inoculated plants. A possible explanation is that the endophyte stimulated Se volatilization from the plant, resulting in lower plant Se levels. Many bacteria (and plants) can volatilize Se, and bacteria have been reported previously to affect plant Se volatilization (de Souza et al., 1999). For a fuller discussion on the microbial community structure of Se hyperaccumulators and non-hyperaccumulators the reader is referred to a related paper recently published (Sura-de

Jong et al., 2015). In future studies it will be interesting to further investigate the nature of these plant-microbe interactions and their ecological implications as well as their applications in agriculture and bioremediation.

3.5. Conclusions

A microbial endophyte isolated from the Se hyperaccumulator plant *Stanleya pinnata* has shown remarkable selenium tolerance. Multi-locus sequence analysis using 16S rRNA and housekeeping genes (*gyrB, rpoB, rpoD*) enabled the phylogenetic classification of the isolateunder investigationto the level of subspecies. The isolate was identified as belonging to *Pseudomonas moraviensis* and named *P. moraviensis* stanleyae. Under aerobic conditions, the isolate could deplete 10 mM of selenite within 48 h. Moreover, it was further shown to grow and produce red elemental selenium in concentrated solutions up to 120 mM of selenite. The reduction of selenite seems to be a detoxification mechanism and the formation of elemental Se was linked to biotic conversion taking place in the stationary phase. During stationary phase, Se(0) nanoparticles were found attached on the surface of the cells and displayed a diameter around 100 nm. Red Se(0) was also produced under anaerobic environments but the growth was potentially limited by the lack of an alternative electron acceptor. Selenate could not be reduced under both aerobic and anaerobic conditions but it did not elicit a toxic effect on the isolate. The fast aerobic growth and the ability to deplete high levels of selenite exhibited by the novel strain of *P. moraviensis* reported herein hold promise as an alternative to the more costly and energy-intensive anaerobic bioreactors or physical-chemical treatment processes. As a follow-up study, it would be interesting to investigate the performance of the newly identified species of *P. moraviensis* on treating a Se-laden wastewater.

3.6. References

Ayano, H., Miyake, M., Terasawa, K., Kuroda, M., Soda, S., Sakaguchi, T., & Ike, M. (2014). Isolation of a selenite-reducing and cadmium-resistant bacterium Pseudomonas sp. strain RB for microbial synthesis of CdSe nanoparticles. *J. Biosci. Bioeng. 117*, 576-581.

Bruins, M.R., Kapil, S., & Oehme, F.W. (2000). Microbial resistance to metals in the environment. *Ecotoxicol. Environ. Saf. 45*, 198-207.

Cappa, J.J., Cappa, P.J., El Mehdawi, A.F., McAleer, J.M., Simmons, M.P., & Pilon-Smits, E.A.H. (2014). Characterization of selenium and sulfur accumulation in *Stanleya* (Brassicaceae). A field survey and common-garden experiment. *Am. J. Bot. 101*, 830-839.

Chapman, P.M., Adams, W.J., Brooks, M., Delos, C.G., Luoma, S.N., Maher, W.A., Ohlendorf, H.M., Presser, T.S., & Shaw, P. (2010). *Ecological assessment of selenium in the aquatic environment*. SETAC Press, Pensacola, FL.

de Souza, M.P., Chu, D., May, Z., Zayed, A.M., Ruzin, S.E., Schichnes, D., and Terry, N. (1999). Rhizosphere bacteria enhance selenium accumulation and volatilization by Indian mustard. *Plant Physiol. 199*, 565-573.

di Gregorio, S., Lampis, S., & Vallini, G. (2005). Selenite precipitation by a rhizospheric strain of *Stenotrophomonas* sp. isolated from the root system of *Astragalus bisulcatus*: a biotechnological perspective. *Environ. Int. 31*, 233-241.

El Mehdawi, A.F., & Pilon-Smits, E.A.H. (2012). Ecological aspects of plant selenium hyperaccumulation. *Plant. Biol. 14*, 1-10.

Fassel, V. A. (1978). Quantitative elemental analyses by plasma emission spectroscopy. *Science 202*, 183-191.

Freeman, J.L., Zhang, L.H., Marcus, M.A., Fakra, S., & Pilon-Smits, E.A.H. (2006). Spatial imaging, speciation and quantification of selenium in the hyperaccumulator plants *Astragalus bisulcatus* and *Stanleya pinnata*. *Plant Physiol. 142*, 124-134.

Galeas, M.L., Zhang, L.H., Freeman, J.L., Wegner, M., & Pilon-Smits, E.A.H. (2007). Seasonal fluctuations of selenium and sulfur accumulation in selenium hyperaccumulators and related non-accumulators. *New Phytol. 173*, 517-525.

Hansen, D., Duda, P., Zayed, A.M., & Terry, N. (1998). Selenium removal by constructed wetlands: role of biological volatilization. *Environ. Sci. Technol. 32*, 591–597.

Hunter, W.J., & Manter, D.K. (2011). *Pseudomonas seleniipraecipitatus* sp. nov.: A selenite reducing γ-Proteobacteria isolated from soil. *Curr. Microbiol. 62*, 565-569.

Ike, M., Takahashim, K., Fujitam, T., Kashiwa, M., & Fujita, M. (2000). Selenate reduction by bacteria isolated from aquatic environment free from selenium contamination. *Water Res. 34*, 3019-3025.

Johnson, J.A., Sabounghi, M.L., Thiyagarajan, P., Csenncsits, R., & Meisel, D. (1999). Selenium nanoparticles: a small-angle neutron scattering study. *J. Phys. Chem. B 103*, 59-63.

Kessi, J., Ramuz, M., Wehrli, E., Spycher, M., & Bachofen, R. (1999). Reduction of selenite and detoxification of elemental selenium by the phototrophic bacterium *Rhodospirillum rubrum*. *Appl. Environ. Microbiol. 65*, 4734-4740.

Kumar, A., Sevonkaev, I., & Goia, D.V. (2014). Synthesis of selenium particles with various morphologies. *J. Colloid. Interf. Sci. 416*, 119-123.

Kuroda, M., Notaguchi, E., Sato, A., Yoshioka, M., Hasegawa, A., Kagami, T., Narita, T., Yamashita, M., Sei, K., Soda, S., & Ike, M. (2011). Characterization of *Pseudomonas stutzeri* NT-I capable of removing soluble selenium from the aqueous phase under aerobic conditions. *J. Biosci. Bioeng. 122*, 259-264.

Lane, D.J. (1991). 16S/23S rRNA sequencing.*In*E. Stackebrandt & M. Goodfellow (ed.), *Nucleic acid techniques in bacterial systematics*. John Wiley and Sons, New York.

Lampis, S., Zonaro, E., Bertolini, C., Bernardi, P., Butler, C.S., & Vallini, G. (2014). Delayed formation of zero-valent selenium nanoparticles by *Bacillus mycoides* SelTE01 as a consequence of selenite reduction under aerobic conditions. *Microb. Cell Fact. 13*, 35.

Lemly, A.D. (2004). Aquatic selenium pollution is a global environmental safety issue. *Ecotox. Environ. Safe. 59*, 44-56.

Lenz, M., & Lens, P.N.L. (2009). The essential toxin: The changing perception of selenium in environmental sciences. *Sci. Total. Environ. 407*, 3620-3633.

Mishra, R.R., Prajapati, S., Das, J., Dangar, T.K., Das, N., & Thatoi, H. (2011). Reduction of selenite to red elemental selenium by moderately halotolerant *Bacillus megaterium*strains isolated from Bhitarkanika mangrove soil and characterization of reduced product. *Chemosphere 84*, 1231-1237.

Mulet, M., Bennasar, A., Lalucat, J., & García-Valdes, E. (2009). An rpoD-based PCR procedure for the identification of Pseudomonas species and for their detection in environmental samples. *Mol. Cell. Probes 23*, 140-147.

Mulet, M., Lalucat, J., & Garcia-Valdes, E. (2010). DNA sequence-based analysis of the *Pseudomonas* species. *Environ. Microbiol. 12*, 1513-1530.

North American Metal Council (NAMC) (2010). Review of available technologies for removal of selenium from water. Available at: http://www.namc.org/docs/00062756.PDF.

Noblitt, S.D., & Henry, C.S. (2008). Improving the compatibility of contact conductivity detection with microchip electrophoresis using a bubble cell. *Anal. Chem. 80*, 7624-7630.

Noblitt, S.D., Staicu, L.C., Ackerson, C.J., & Henry, C.S. (2014). Sensitive, selective analysis of selenium oxoanions using microchip electrophoresis with contact conductivity detection. *Anal. Chem. 86*(16), 8425-8432.

Oremland, R.S., Mitchell, H.J., Switzer Blum, J., Langley, S., Beveridge, T.J., Ajayan, P.M., Sutto, T., Ellis, A., & Curran, S. (2004). Structural and spectral features of selenium nanospheres produced by Se respiring bacteria. *Appl. Environ. Microbiol. 70*, 52-60.

Palleroni, N.J. (2005). Genus 1. *Pseudomonas*. In: Brenner, D.J., Krieg, N.R., Staley, T.J., & Garrity, G.M. (eds.) *Bergey's manual of systematic bacteriology*. The *Proteobacteria* part B, The Gammaproteobacteria. 2nd Edition. Springer, New York, USA, pp. 323-379.

Pearce, C.I., Coker, V.S., Charnock, J.M., Pattrick, R.A.D., Mosselmans, J.F.W., Law, N., Beveridge, T.J., & Lloyd, J.R. (2008). Microbial manufacture of chalcogenide-based nanoparticles via the reduction of selenite using *Veillonellaatypica*: an *in situ* EXAFS study. *Nanotechnology 19*, 155603.

Peix, A., Ramirez-Bahena, M.H., & Velazquez, E. (2009). Historical evolution and current status of the taxonomy of genus *Pseudomonas*. *Infect. Genet. Evol. 9*, 1132-1147.

Rajkumar, M., Sandhya, S., Prasad, M.N.V., & Freitas, H. (2012). Perspectives of plant-associated microbes in heavy metal phytoremediation. *Biotechnol. Adv. 30*, 1562-1574.

Simmons, D.B., & Wallschlaeger, D. (2005). A critical review of the biogeochemistry and ecotoxicology of selenium in lotic and lentic environments. *Environ. Toxicol. Chem. 24*, 1331-1343.

Sobolewski, A. (2013). Evaluation of treatment options to reduce water-borne selenium at coal mines in West-Central Alberta. Microbial Technologies, Inc. Available from: http://environment.gov.ab.ca/info/library/7766.pdf.

Spiers, A.J., Buckling, A., & Rainey, P.B. (2000). The causes of *Pseudomonas* diversity. *Microbiology 146*, 2345-2350.

Staicu, L.C., van Hullebusch, E.D., Lens, P.N.L., Pilon-Smits, E.A.H. & Oturan, M.A. (2015). Electrocoagulation of colloidal biogenic selenium. *Environ. Sci. Pollut. Res. Int.* 22, 3127-3137.

Sura-de Jong, M., Reynolds, J., Richterova, K., Musilova, L., Staicu, L.C., Chocholata, I., Cappa, J.J., Taghavi, S., van der Lelie, D., Frantik, T., Dolinova, I., Strejcek, M., Cochran, A.T., Lovecka, P. & Pilon-Smits, E.A.H. (2015). Selenium hyperaccumulators harbor a diverse endophytic bacterial community characterized by high selenium tolerance and growth promoting properties. *Front. Plant Sci. 6*, 113.

Switzer Blum, J., Bindi, A.B., Buzzelli, J., Stolz, J.F., & Oremland, R.S. (1998). *Bacillus arsenoselenatis* sp. nov., and *Bacillus selenitireducens* sp. nov.: two haloalkaliphiles from Mono Lake, California, which respire oxyanions of selenium and arsenic. *Arch. Microbiol. 171*, 19-30.

Tayeb, L., Ageron, E., Grimont, F., & Grimont, P.A.D. (2005). Molecular phylogeny of the genus Pseudomonas based on rpoB sequences and application for the identification of isolates. *Res. Microbiol. 156*, 763-773.

Terry, N., Zayed, A.M., de Souza, M.P., & Tarun, A.S. (2000). Selenium in higher plants. *Annu. Rev. Plant Physiol. Plant Mol. Biol. 51*, 401-432.

Tvrzova, L., Schumann, P., Sproer, C., Sedlacek, I., Pacova, Z., Sedo, O., Zdrahal, Z., Steffen, M., & Lang, E. (2006). *Pseudomonas moraviensis* sp. nov. and *Pseudomonas vranovensis* sp. nov., soil bacteria isolated on nitroaromatic compounds, and emended description of *Pseudomonas asplenii*. *Int. J. Syst. Evol. Microbiol. 56*, 2657-2663.

USEPA (2003). Ground water and drinking water: list of drinking water contaminants and MCLs. Available from: http://www.epa.gov/safewater/mcl.html#inorganic

White, C., Sayer, J.A., & Badd, G.M. (1997). Microbial solubilization and immobilization of toxic metals: key biogeochemical processes for treatment of contamination. *FEMS Microbiol. Rev. 20*, 503-516.

Yamamoto, S., Kasai, H., Arnold, D.L., Jackson, R.W., Vivian, A., & Harayama, S. (2000). Phylogeny of the genus Pseudomonas: intrageneric structure reconstructed from the nucleotide sequences of gyrB and rpoD genes. *Microbiology 146*, 2385-2394.

Zago, A., & Chugani, S. (2009). *Pseudomonas*. In: Moselio Schaechter (ed.) *Encyclopedia of Microbiology*. 3rd edition, Vol. 1. Bacteria. Elsevier, San Diego, USA, pp. 245-326.

Zarcinas, B. A., Cartwright, B. & Spouncer, L. R. (1987). Nitric acid digestion and multielement analysis of plant material by inductively coupled plasma spectrometry. *Commun. Soil Plant Anal. 18*, 131-146.

CHAPTER 4

Electrocoagulation of colloidal biogenic selenium

This chapter was published as:

Staicu, L.C., van Hullebusch, E.D., Lens, P.N.L., Pilon-Smits, E.A.H., & Oturan, M.A. (2015). Electrocoagulation of colloidal biogenic selenium. *Environ. Sci. Pollut. Res. Int.* 22(4), 3127-3137.

Chapter 4. Electrocoagulation of colloidal biogenic selenium

Abstract

Colloidal elemental selenium, Se(0), adversely affects membrane separation processes and aquatic ecosystems. As a solution to this problem we investigated for the first time the removal potential of Se(0) by electrocoagulation process. Colloidal Se(0) was produced by a strain of *Pseudomonas moraviensis* and showed limited gravitational settling. Therefore, iron (Fe) and aluminum (Al) sacrificial electrodes were used in a batch reactor under galvanostatic conditions. The best Se(0) turbidity removal (97%) was achieved using iron electrodes at 200 mA. Aluminum electrodes removed 96% of colloidal Se(0) only at a higher current intensity (300 mA). At the best Se(0) removal efficiency, electrocoagulation using Fe electrode removed 93% of the Se concentration, whereas with Al electrodes the Se removal efficiency reached only 54%. Due to the less compact nature of the Al flocs, the Se-Al sediment was three times more voluminous than the Se-Fe sediment. The TCLP test showed that the Fe-Se sediment released Se below the regulatory level (1 mg L^{-1}), whereas the Se concentration leached from the Al-Se sediment exceeded the limit by about 20 times. This might be related to the mineralogical nature of the sediments. Electron scanning micrographs showed Fe-Se sediments with a reticular structure, whereas the Al-Se sediments lacked an organized structure. Overall, the results obtained showed that the use of Fe electrodes as soluble anode in electrocoagulation constitutes a better option than Al electrodes for the electrochemical sedimentation of colloidal Se(0).

Keywords: Elemental selenium; Colloids; Electrocoagulation; Aluminum electrodes; Iron electrodes; Toxicity Chaacteristic Laching Procedure (TCLP).

4.1. Introduction

Selenium (Se) has a complex biogeochemistry with both abiotic and biotic reactions involved in its cycling through different compartments of the environment. The two most oxidized forms, or oxyanions, namely selenite (Se[IV], SeO$_3^{2-}$) and selenate (Se[VI], SeO$_4^{2-}$), are water-soluble, bioavailable and toxic (Simmons & Wallschlaeger, 2005). During the mid-1970s, Lake Belews in North Carolina was affected by Se oxyanions released by a coal-fired power plant, which resulted in the massive die-off of the local fish populations: 19 out of 20 species were eliminated (Lemly, 2002). In the early 1980s, the Se-laden agricultural drain water discharged in the Kesterson Reservoir, California, severely affected the migratory bird populations and triggered environmental actions (Ohlendorf, 1989).

In contrast to its soluble oxyanions, elemental Se, Se(0), is particulate and less bioavailable (Lenz & Lens 2009). However, when released into surface waters, Se(0) has been reported to adversely affect bivalve mollusks (Luoma et al., 1992; Schlekat et al., 2000) and to be oxidized to Se oxyanions (Zhang et al., 2004). Because the filter-feeding mollusks are situated at the bottom of the trophic network, their Se content is biomagnified in the higher trophic levels (Chapman et al., 2010). The impact of colloidal elemental Se(0) on bivalve mollusks and trophic networks was investigated extensively in the San Francisco Bay area (Schlekat et al., 2000; Purkerson et al., 2003; Presser & Luoma, 2006; USEPA, 2010). To complicate matters further, biogenic Se(0) exhibits colloidal properties that make its separation from aqueous solution problematic (Buchs et al., 2013; Staicu et al., 2015a).

Major sources of wastewaters containing selenium oxyanions are oil refining industry, coal combustion and metal refining (Lemly, 2004; USEPA, 2010). The biological treatment of these wastewaters produces different concentration levels of colloidal Se(0) as a function of the initial Se content and the Se conversion rate. Removal of colloidal Se(0) from the bioreactor effluent is necessary to reduce its environmental load and the negative impact exerted on aquatic ecosystems.

During the last decades, electrocoagulation (EC) has gained recognition as a powerful water treatment technology (Holt et al., 2005) to remove colloidal species from water. In EC, the electrical current is applied between two electrodes (including a sacrificial anode) immersed in the (waste)water to be treated. Applying a current across electrodes creates an electrical field, causes the electrolysis of water and the dissolution of sacrificial anode to form a coagulant. The coagulants are thus electrogenerated *in situ* and in a continuous manner during the application of the process.

The main reactions taking place at the anode and cathode in an electrolytic cell are the following (Mollah et al., 2004):

At the anode:

$$M\ (s) \rightarrow M^{n+}\ (aq) + ne \qquad (4.1)$$

$$2H_2O \rightarrow O_2\ (g) + 4H^+ + 4e^- \qquad (4.2)$$

At the cathode:

$$2H_2O + 2e^- \rightarrow H_2\ (g) + 2OH^- \qquad (4.3)$$

In the bulk solution:

$$M^{n+} + nOH^- \rightarrow M(OH)n\ (s) \qquad (4.4)$$

where M is the metal (e.g. Al, Fe) in its elemental form (zero valence state), M^{n+} is the oxidized metal (n = 2, 3), ne^- represents the number of electrons transferred. Due to their proven efficiency and affordable price, Al and Fe electrodes are frequently employed in EC processes. In the case of Fe anode, the Fe^{2+} ions are oxidized by dissolved O_2 to form Fe^{3+}. The release of polyvalent cations neutralizes the negatively-charged colloidal particles leading to their destabilization and aggregation. In addition, the presence of an electrical field enhances the collision probability and therefore the efficiency of the coagulation process.

Equations 4.1 and 4.4 explain the formation of metallic cations and their hydroxides that react to form hydroxo monomeric and polymeric species. Metal hydroxides have large surface areas involved in the adsorption of soluble and colloidal particles (Mollah et al., 2001). In addition, gas bubbles in the form of oxygen (4.2) and hydrogen (4.3) are generated by both electrodes. The gas bubble formation can negatively impact the settling efficiency of the contaminant-metal hydroxide flocs. In order to minimize this effect, lower currents and vertical electrode configurations are suggested (Heidmann & Calmano, 2010).

Various electrocoagulation studies have focused on treating toxic metals (Vasudevan et al., 2009; Akbal & Camci, 2011; Al Aji et al., 2012; Mello Ferreira et al., 2013; Öncel et al., 2013), turbidity (Trompette et al., 2008; Gamage & Chellam, 2011; Khandegar & Saroha, 2013), microorganisms (Zhu et al., 2006) or complex organic matrices (Un et al., 2009). Kabdasli et al., (2012) and Vasudevan & Oturan (2014) provide an extensive and in-depth analysis of the wastewater types treated by electrocoagulation. To our best knowledge, there is no study reported in the literature about the elimination of colloidal Se(0) from water by EC. Therefore the aim of this study was to investigate the viability for the removal of colloidal biogenic Se(0) by EC. Colloidal Se(0) was produced by a strain of *Pseudomonas moraviensis* (Staicu et al., 2015b). Fe and Al electrodes were used as sacrificial anodes. The colloidal Se(0) removal capacity of the system was evaluated using different operating parameters, including current density and electrode type. Coagulation efficiency was determined by turbidity measurements. Furthermore, the residual metal content of the supernatant was determined. The sediments produced at the end of the EC were analyzed in terms of structure and metal leaching behavior.

4.2. Materials and Methods

4.2.1. Reagents and electrodes

For bacterial growth, King B (KB) medium was prepared as described by King et al., (1954). Sodium selenite, Na_2SeO_3, >99%, was purchased from Sigma Aldrich. Coagulant reagent was generated by the electrodissolution of Al (99% purity) and Fe (99.5% purity) electrodes both from Goodfellow Ltd., UK.

All solutions were prepared using deionized water. Before and after each experiment, the electrodes were degreased by wiping with an acetone-soaked tissue, abraded with sand paper and rinsed with ultrapure water to remove any impurities and oxide layers (Heidmann & Calmano, 2010).

4.2.2. Biogenic Se(0) production and solution preparation

Biogenic elemental Se, Se(0), was produced aerobically by a strain of *Pseudomonas moraviensis*, isolated from the roots of *Stanleya pinnata* (Fam. *Brassicaceae*), a model Se hyperaccumulator species (Staicu et al., 2015b). The KB medium was supplemented with 10 mM Na_2SeO_3 from a filter-sterilized 1 mol L^{-1} stock solution and with 1% (v/v) of *P. moraviensis* inoculum sampled during the mid-logarithmic growth phase. The incubation was performed under aerobic conditions at 28 °C; initial pH, pH_0 = 7.5; and under shaking at 160 rpm. After 24 hours, the incubation turned red, indicative of SeO_3^{2-} reduction to red elemental Se. Elemental Se was harvested by centrifugation (at 3,200 x *g* for 10 min), washed twice with deionized water and re-suspended in 42 mM NaCl solution (Canizares et al., 2007). The electrolyte addition corresponded to a conductivity of around 4.5 mS cm^{-1}. The turbidity of the solution was adjusted by adding biogenic Se(0) or NaCl solution until the desired target value was reached. Due to the colloidal nature of Se(0) the targeted turbidity value of the suspension was set within ± 5%. NaCl acts as supporting electrolyte.

4.2.3. Electrocoagulation set-up

The coagulating agents were electrogenerated using sacrificial metallic anodes of Al and Fe, respectively (Figure 4.1). The electrochemical experiments were conducted under galvanostatic conditions, i.e. the current was set and the potential adjusted its value as a function of system's resistance. The electrocoagulation experiments were carried out in batch mode in a 500 mL single compartment electrochemical cell containing colloidal Se(0) suspension and two electrodes (100 mm height x 50 mm width x 1 mm thick in dimension, with 6 cm depth in solution). Both the anode and the cathode are consisting of same metal (Al or Fe). The electrodes were connected in a monopolar mode and were placed vertically and parallel to each other. This electrode configuration was chosen to minimize the flotation effect of the hydrogen and oxygen gas bubble evolution exerted on the colloidal Se(0) suspension. The conversion between the applied current (I) and the current density J (J = I/A, with A, surface area) is presented in Table 4.1. For ease of comparison between datasets we employed the current values for generating the graphs.

Table 4.1. Summary of the EC results using Al and Fe sacrificial electrodes.

Fe electrodes

I (mA)	j (mA cm^{-2})	U (V)	Removal efficiency (%)	Residual turbidity (NTU)	Fe$_{theo}$ (g L^{-1})	E (kWh m^{-3})	C$_{electrode}$ (kg m^{-3})	M$_{sediment}$ (kg m^{-3})
50	0.83	1.4	69	157	0.1	0.14	0.104	17.4
100	1.67	2.3	81	95	0. 21	0.46	0.208	39.2
200	3.33	3.9	97	16	0.42	1.56	0.417	61.6
300	5.00	5.1	97	18	0.63	3.06	0.625	71.8
500	8.33	7.9	96	20	1.04	7.9	1.042	84.5

Notes: Fe theoretical, Fe$_{theo}$ (mg L^{-1}); Electrical energy consumption, E (kWh m^{-3}); Electrode consumption, C$_{electrode}$ (kg m^{-3}); Mass of sediment, M$_{sediment}$ (kg m^{-3}) were determined after 60 min of electrolysis.

Al electrodes

I (mA)	j (mA cm^{-2})	U(V)	Removal efficiency (%)	Residual turbidity (NTU)	Al$_{theo}$ (g L^{-1})	E (kWh m^{-3})	C$_{electrode}$ (kg m^{-3})	M$_{sediment}$ (kg m^{-3})
50	0.83	1.9	71	151	0.03	0.19	0.034	32
100	1.67	2.6	86	86	0.07	0.52	0.067	76
200	3.33	4.2	88	61	0.13	1.68	0.134	150
300	5.00	5.7	96	22	0.2	3.42	0.201	212
500	8.33	8.5	95	25	0.33	8.5	0.336	376

Notes: Al theoretical, Al$_{theo}$ (mg L^{-1}); Electrical energy consumption, E (kWh m^{-3}); Electrode consumption, C$_{electrode}$ (kg m^{-3}); Mass of sediment, M$_{sediment}$ (kg m^{-3}) were determined after 60 min of electrolysis.

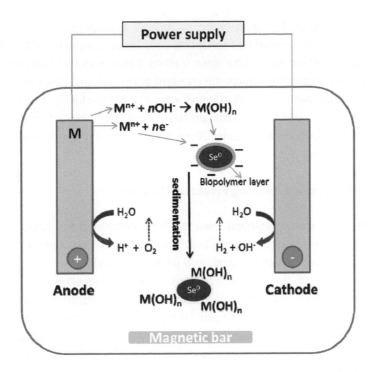

Figure 4.1. Schematic diagram of the electrocoagulation set-up. M = metal (e.g. Al, Fe), n+ = oxidation state; M(OH)$_n$ = metal hydroxides.

In order to improve the mass transfer, the electrochemical reactor was mixed at 300 rpm using a 3 cm magnetic stir bar. The electrodes were immersed in the solution up to a 60 cm^2 active electrode geometric area with a 3 cm electrode gap. The constant agitation produced by the magnetic stir bar ensured the homogeneous mass transfer of the coagulant within the electrochemical reactor and increased the collision frequency of colloidal Se(0) particles with the coagulating agent. Based on a preliminary study (data not shown), we determined the optimal distance to be 3 cm and the optimal stirring speed to be 300 rpm. The distance between electrodes is a critical parameter in the electrochemical cell design since a suboptimal electrode gap increases the IR-drop leading to higher energy consumption (Mollah et al., 2004). The results are presented as average values of three independent experiments (triplicates, $n = 3$) unless otherwise stated. When the standard deviation values were smaller than 5%, the error bars were not represented. All data was analyzed by using the data analysis software SigmaPlot 12.0v.

4.2.4. Electrocoagulant generation

In order to determine the amount of Al and Fe electrogenerated, separate experiments were performed. The Se(0)-free solution contained 42 mM NaCl and the sampling was done with the same frequency as the EC Se(0) treatment protocol. All samples were acidified with

65% HNO_3 and stored at 4 °C until elemental analysis was performed on a PerkinElmer Optima 8300 Inductively Coupled Plasma-Optical Emission Spectrometer (ICP-OES). Residual Se, Fe and Al were determined by the same method. Calibration standards were prepared by stepwise dilution of a multi-element ICP standard solution (Merck, Darmstadt, Germany). The wavelengths employed were 196.026 nm for Se, 238.204 nm for Fe, and 396.153 nm for Al, respectively. At the end of the sedimentation stage, the supernatants were carefully siphoned from the sedimentation cones. Both the supernatant and the sediments were collected for further analyses.

4.2.5. Toxicity Characteristic Leaching Procedure (TCLP) test

The sediment samples were analyzed using the TCLP performed according to the testing guidelines specified by the United States Environmental Protection Agency (USEPA, 1999). The sediment was mixed in a glacial acetic acid (of 99.5% assay) solution at 1:20 with a final pH of 2.88 ± 0.05. The leachate mixture was sealed in the extraction vessel (plastic centrifuge tubes, 29 x 115 mm, 50 mL) and tumbled for 20 hours using a Grant Bio PTR-30 360° Vertical Multi-Function Rotator to simulate an extended leaching time in the ground. The vertical rotation speed employed was 30 rpm. After 20 h, the samples were filtered gravitationally through Whatman glass microfiber filters, Grade GF/F (0.7 μm cut-off) and the filtrate analyzed by ICP-OES.

4.2.6. Analytical methods

Turbidity wasmeasured using a HACH 2100P ISO turbidimeter (Hach 2100P ISO) containing a T860 nm LED lamp and was expressed in Nephelometric Turbidity Units (NTU). 15 mL aliquots were sampled according to the manufacturer's instructions. Electrophoretic mobility measurements were performed on a Zetasizer Nano ZS (Malvern Instrument Ltd., Worchestershire, UK) using a laser beam at 633 nm and a scattering angle of 173° at 25 °C according to the manufacturer's instructions. The volume of settled Se(0) was measured in standard 1,000 mL Imhoff graduated cones (USEPA, 1999).

Environmental scanning electron microscopy (ESEM, ELECTROSCAN E3, Hillsboro, OR, USA) was used for the observation of the microstructure of Se-Fe and Se-Al sediments. ESEM allows the examination of wet specimens without sample preparation (Donald, 2003). The samples were observed at 25 kV. X-ray diffraction (XRD) analysis was performed on a Bruker D8 Advance diffractometer (Karlsruhe, Germany) equipped with an energy dispersion Sol-X detector with copper radiation (CuKα, λ = 0.15406 nm). The acquisition was recorded between 10° and 80°, with a 0.02° scan step and 1 s step time.

pH was measured by an EUTECH 1500 pH meter. Conductivity measurements were performed on a Radiometer Analytical MeterLab CDM 230. The electrical current was applied and the evolution of the current and voltage was monitored using a HAMEG Triple Power HM7042-5 (Mainhausen, Germany).

4.2.7. Calculations

The specific electrical energy consumption E (kWh m^{-3}) for turbidity removal was calculated as follows (Heidmann & Calmano, 2010):

$$E = \frac{U \cdot I \cdot t}{V \cdot 1000} \tag{4.5}$$

Where U is the required voltage (V), I, the applied current (A), t, electrolysis time (h) and V, the volume of the treated solution (m^3).

The maximum possible mass of Al and Fe electrochemically generated from sacrificial anodes for a particular electrical current was calculated using Faraday's law of electrolysis (Mechelhoff et al., 2013):

$$m = \frac{j \cdot Ael \cdot M \cdot t}{V \cdot z \cdot F} \tag{4.6}$$

where m is the mass of the anode material dissolved (g), j the current density (A m^{-2}), Ael the active electrode area (m^2), M the molar mass of the anode material (g mol^{-1}), t electrolysis time (s), V volume of the reactor (m^3), z the number of electrons transferred, and F the Faraday's constant (96,485 C mol^{-1}). The cathode dissolution was not considered.

4.3. Results and Discussion

4.3.1. Characterization of the biogenic Se(0) suspension

The properties of the biogenic red Se(0) solution (Figure 4.2b) are summarized in Table 4.2. The suspension is characterized by high turbidity and neutral pH. The Se(0) particles exhibit a negative surface charge of around -20 mV.

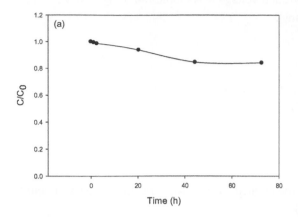
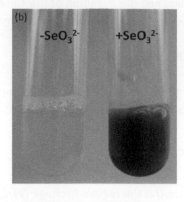

Figure 4.2. (a) Colloidal stability of biogenic red Se(0) produced by *P. moraviensis*.Evolution of the ratio C/C_0 in function of treatment time with C_0 *and C* the turbidity in NTU unit at the initial and a given time, respectively and (b) Biogenic red Se(0) produced by *P. moraviensis* and the control sample (KB medium).

The colloidal stability of biogenic Se(0) is presented in Figure 4.2a. Biogenic Se(0) displayed limited sedimentation during the 4-day study interval, from 500 NTU to 420 NTU, corresponding to a normalized removal of 0.16. Note that turbidity was normalized against its initial value using a scale from 0 to 1 (*1 represents no sedimentation, 0 represents total sedimentation of the colloidal system*). The colloidal stability of Se(0) is related to the negatively-charged biopolymers that are coating the biogenic selenium particles (Dobias et al., 2011; Lenz et al., 2011; Buchs et al., 2013).

Table 4.2. Properties of biogenic Se(0) solution produced by a *P. moraviensis* strain (Growth conditions: 28 °C; 160 rpm; pH_0 = 7.5; aerobic; incubation time, 24 h).

Parameter	Value
Turbidity (NTU)	500 ± 30
Se (mg L^{-1})	310 ± 12
Color	Red
pH	7.0 ± 0.2
Conductivity (mS cm^{-1})	4.5 ± 0.2
Zeta potential (mV)	-20 ± 2

4.3.2. Electrodissolution of the Al and Fe electrodes

Figure 4.3 compares the variation the concentrations of electrogenerated Al and Fe measured with respect to the electrical charge passed. For both electrodes, the metal concentration increases linearly with the electrical charge. The measured values exceed the predicted values calculated from Eq. 4.6, i.e. a super-faradaic yield is obtained. In the case of Al, the difference is negligible, whereas the Fe anode exhibits 1.3 times higher measured concentrations than the theoretically expected ones. Even if Fe has a higher redox potential (-0.44 V) than Al (-1.67 V), the formation of an Al oxide layer during electrolysis coats the electrode surface, thus decreasing its corrosion potential (Roberge, 2008).

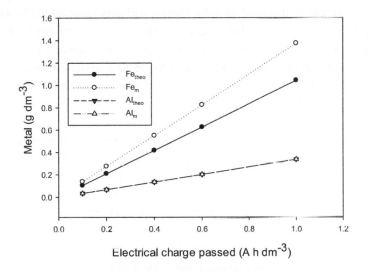

Figure 4.3. Variation of theoretical and measured Al and Fe with the electrical charge passed during EC. Operatingconditions: temperature, 20 °C; volume, 0.5 L; supporting media, 42 mmol L^{-1} of NaCl; initial pH, pH_0 = 7.0; 300 rpm; electrolysis time, 60 min. Fe_{theo} and Al_{theo} refer to the theoretical metal dissolved according to Faraday's law (Eq. 4.6)). Fe_m and Al_m are the measured values with charge passed.

The super-faradaic yield exhibited by the Fe electrodes could be explained by the difference between the electrode's geometric area and the actual area that takes into account the surface roughness. Electrocoagulation is corroding the sacrificial anodes, thus increasing their surface area as a function of the anode material, current applied and the corrosiveness of the solution (Roberge, 2008). An alternative explanation considers the additional chemical dissolution of the Fe anode (Canizares et al., 2005) due to corrosion. In addition, Picard et al., (2000) reported cathodic dissolution as a consequence of the chemical attack by the hydroxyl ions released during water electrolysis. However, they used a very high conductivity (0.6 mol L^{-1} NaCl) and the currents used were 200 times higher than those applied in the current study.

4.3.3. Treatment efficiency of electrocoagulation

Figure 4.4a compares the turbidity removal using Fe and Al anodes as a function of the current intensity. For the first two current values applied (50 and 100 mA), the Al electrodes produced a slightly higher turbidity removal than the Fe electrodes. At 50 and 100 mA, the Al electrodes removed 71% and 86% of the initial turbidity, whereas the Fe electrodes achieved 69% and 81%, respectively. In contrast, Fe electrodes lead to a better turbidity removal at higher applied current intensity values. For example, when increasing the current to 200 mA, the Fe electrodes produced the highest turbidity removal (97%). At the same current, Al generated only 88% turbidity removal. Above 200 mA, the electrodes displayed different trends. While turbidity removal by the Fe electrodes remained almost constant at around 97% for 300 mA and 500 mA, Al showed the highest performance at 300 mA, with 96% removal efficiency, followed by 95% removal at 500 mA. Overall, these results indicate a plateau above 200 mA for the Fe electrodes and above 300 mA for the Al electrodes.

Because colloids and electrically-charged ions are held in suspension due to the electrostatic repulsion forces, the presence of counter ions brings about neutralization of the electric charge and diminishes their colloidal stability (Khandegar & Saroha, 2013). On the other hand, when liberated in the bulk solution, the metal cations hydrolyze spontaneously by forming a series of metastable hydrolysis products that transit towards thermodynamically stable metal hydroxides (Richens, 1997). Destabilized colloidal particles are adsorbed onto metal (oxy)hydroxides or followed by precipitation (Hanai & Hasar, 2011). As a consequence of these mechanisms, the colloids aggregate and settle down.

The relation between the current applied and the turbidity removal was not linear (Figure 4.4a). This indicates a decrease of current efficiency during the electrolysis time, either due to parasitic reactions that are enhanced with the increase of the applied current or to the formation of a passivation layer. Since the temperature measured during the experiments increased only marginally (less than 0.5 °C) compared to the beginning of the experiment, the Joule effect (i.e. heating induced by the increase of the applied current) seems not to play an important role (Kabdasli et al., 2012). Passivation of the electrode surface has been shown to affect the performance of the process (Mouedhen et al., 2008; Lakshmanan et al., 2009). Because the highest removal efficiencies were recorded at the beginning of the plateau, no further experiments were performed at currents above 500 mA.

A visual comparison between the treatment effectiveness of Al and Fe electrodes at 200 mA is provided in Figure 4S (page 100).

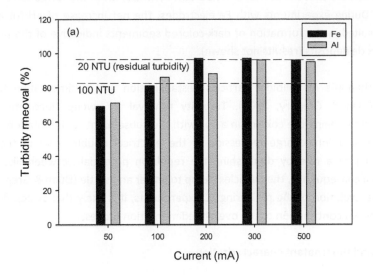

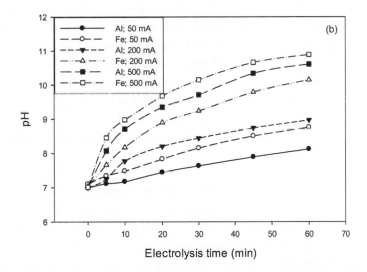

Figure 4.4. (a) Turbidity removal and (b) pH evolution for Fe and Al electrodes. Note that the dashed lines in panel (a) represent residual turbidity (NTU). Operating conditions: temperature, 20 °C; volume, 0.5 L; supporting electrolyte, 42 mmol L^{-1} of NaCl; pH$_0$ = 7.0; 300 rpm; electrolysis time, 60 min.

pH is an important factor that influences the speciation of Fe and Al during the process (Mollah et al., 2001). Figure 4.4b shows the pH evolution during electrolysis time as a function of the applied current. All experiments started at a pH value of 7.0 ± 0.2. For each current intensity tested, the Fe electrodes induced a higher pH increase than the Al ones. It is important to note that the pH has an important contribution to the formation of Fe and Al hydroxides. During experiments with Fe electrodes, the net increase of pH for all current densities resulted in the formation of dark-colored sediments indicative of the presence of ferric hydroxide deposits (results not shown).

Several mechanisms of colloidal particles destabilization and sedimentation have been proposed (Duan & Gregory, 2003). Turbidity removal by sweep flocculation acts by entrapping and bridging the colloids in a floc with the subsequent sedimentation (Mollah et al., 2004). In addition, charge repression of the electrical double layer of the colloidal particles can play a role by diminishing the repulsion potential between likely-charged particles. As a consequence, the particles clump together and settle (Duan & Gregory, 2003). Based on the evolution of the pH during EC experiments, it is likely that sweep flocculation had an important contribution to the overall sedimentation process.

4.3.4. TCLP and supernatant characterization

Leaching tests of the electrogenerated Al and Fe sediments were conducted following the TCLP method (USEPA, Method 1311). The sediments generated by 50 mA and 100 mA currents were investigated in terms of the amount of Se, Fe and Al released from the sediment matrix. Figure 4.5a presents the TCLP results of Se-Fe and Se-Al sediment samples. At 50 mA, the Se-Al sediment released 5 times more Se than the Se-Fe sample, 16 mg L^{-1} versus 3 mg L^{-1}. When doubling the applied current, the difference between the Se released by the two sediments increased 22 times: 17.8 mg L^{-1} (for Al treatment) versus 0.8 mg L^{-1} (for Fe treatment). Only the Se concentration of the leachate from the Se-Fe sediment generated at 100 mA complied with the 1 mg L^{-1} USEPA regulatory limit (USEPA, Method 1311).

Al-Se and Fe-Se sediments showed different leaching behavior of Al, Fe and Se, respectively. At 50 mA, the concentration of Al released was two-fold higher than the Fe concentration, whereas at 100 mA the Al concentration was slightly inferior compared to the Fe concentration. Fe and Al are not TCLP-regulated elements (USEPA, Method 1311, Chapter 7). The stability of sediments over time is an open question related to the validity of the TCLP approach as a tool for the sound characterization of the leaching potential of sediments to be landfilled.

Residual turbidity (Figure 4.4a) is an important parameter in wastewater treatment. Low residual turbidity is desirable from a technological and ecotoxicological standpoint. The lowest turbidity achieved by the Fe electrodes was 16 NTU for 200 mA. The Al electrodes achieved a minimum residual turbidity of 22 NTU at 300 mA. While in chemical coagulation, the addition of the coagulant is a discrete event and the system shifts towards a final equilibrium state, in EC the equilibrium is constantly moving. Therefore, trying to determine the turbidity removal kinetics in EC is challenging because of the concomitant formation of metal hydroxides during electrolysis that add to the overall turbidity (Holt et al., 2002).

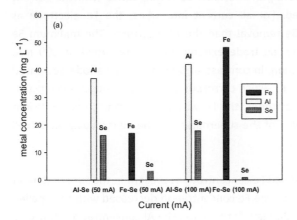

Figure 4.5. (a) TCLP of Se-Al and Se-Fe sediments, (b) Residual Se in the supernatant of Al and Fe electrode experiments Note that "zero" mA corresponds to the initial Se content of the suspension before EC treatment, and (c) Residual Al and Fe from the supernatant of Al and Fe electrode experiments.

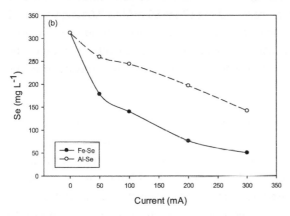

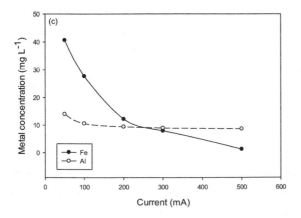

Residual Se (total selenium) remaining in the supernatant after the liquid-solid phase separation is presented in Figure 4.5b. The initial Se concentration of the colloidal Se(0) solution (500 ± 30 NTU) was 310 ± 12 mg L^{-1}. Residual Se (total) varies with the current intensity and the electrode type used. For all current intensities, the Fe electrode as sacrificial anode is more efficient in Se removal than the Al electrode. The minimum Se concentration (141.6 mg L^{-1}) in the Al electrode experiment was recorded at 300 mA, corresponding to around 54% Se reduction. In contrast, at 500 mA the Fe anode decreased the total Se concentration up to 23 mg L^{-1} which corresponds to a Se removal efficiency of around 93%. By comparing the two electrodes, the Fe was almost 6 times more efficient than Al in total Se removal from analysis of the supernatant of the electrocoagulation cell upon termination of the experiment.

Figure 4.5c displays the residual total Al and Fe concentrations present in the supernatant at the end of the sedimentation stage. While the Fe concentration decreased with the applied current, from 41 mg L^{-1} (at 50 mA) to 1.26 mg L^{-1} at 500 mA, the Al concentration showed a limited decrease with the applied current, from 14.2 mg L^{-1} (at 50 mA) to 8.64 mg L^{-1} (at 500 mA). The decrease in Fe concentration can be linked to the decrease in residual Se because Fe (oxy)hydroxides co-precipitate with colloidal Se(0) and is entrapped in the sediment. The higher Se concentration in the Al electrode experiment could also be understood in the same framework.

Secondary pollution refers to pollutants that are not initially present in the wastewater but are introduced during treatment. EC adds the anode-corroded metal to the treated solution. A drawback of using Fe electrodes in EC is related to the residual color (yellowish, green, greenish-black) produced by the dissolution and speciation of Fe (Moreno et al., 2009). It is generally agreed that Al is not required for proper functioning of biological systems (Gensemer & Playle, 1999) and therefore its presence can elicit toxic effects. In contrast, Fe is an essential metabolic component functioning as a cofactor in a wide array of proteins and enzymes (Arredondo & Nunez, 2005). Dong et al. (2014) investigated the Comparative Toxicity Potentials (CTP) of 14 cationic metals, including Al and Fe, in freshwater environments. While Al had one of the highest CTP, Fe ranked among the metals with the lowest CTP. Therefore, from an ecotoxicological standpoint, Fe is a better option for treating colloidal Se(0) suspensions.

4.3.5. Sediment characterization

To clarify the structure of sediments we carried out ESEM micrographs of Se-Al and Se-Fe sediments. Results obtained are depicted on Figures 4.6a and 4.6b. Se-Al does not appear to have an organized structure. Se(0) nanoparticles are clearly visible. In contrast, the Se-Fe sediment shows a reticular structure with no observable Se(0) particles.

X-ray diffraction was performed to investigate the mineralogy of the two types of sediments (Figures 4.6c and 4.6d). Gamage & Chellam (2011) reported that Al sediments generated by EC having an amorphous state and a gelatinous appearance. The same observation was made for our samples (results not shown). The diffractogram of Se-Fe sediments (Figure 4.5d) also indicates an amorphous state. This result is in agreement with that reported by Heidmann & Calmano (2010) concerning iron-based sediments generated by EC.

X-ray diffraction was performed to investigate the mineralogy of the two types of sediments (Figures 4.6c and 4.6d). Gamage & Chellam (2011) reported that Al sediments generated by EC having an amorphous state and a gelatinous appearance. The same observation was made for our samples (results not shown). The diffractogram of Se-Fe sediments (Figure 4.6d) also indicates an amorphous state. This result is in agreement with that reported by Heidmann & Calmano (2010) concerning iron-based sediments generated by EC.

The properties of Se-Fe and Se-Al sediments generated by electrocoagulation using different currents are summarized in Table 4.1. On the other side, the evolution of Se-Fe and Se-Al sediment volume as a function of the applied current can be seen in Figure 4.6e. While the volume of both Al- and Fe- sediments show a positive trend with the increase in current, the Se-Al sediment becomes more voluminous than Se-Fe by a factor of 2 (at 50 mA) to 4.5 (at 500 mA). Because Fe and Al have different densities (7.874 and 2.70 g cm^{-3} for Fe and Al, respectively) this will impact the floc and eventually sediment structure (Lide, 2004; Turchiuli & Fargues, 2004). Moreover, higher currents have been shown to increase the bubble densities and create Al flocs with a less compact structure (Holt et al., 2002). Al flocs produced in EC are fragile and relatively porous, therefore prone to breakage (Harif et al., 2012). Smaller flocs thus created are characterized by a poor settleability and voluminous sediment with a porous structure (Harif & Adin, 2011). The size of the flocs can play a significant role in settling and solid-liquid separation, as well as in the structure of the sediment (Jarvis et al., 2005).

The mass of the Se-Fe and Se-Al sediments is superior to their volume, resulting in a density greater than unity (Table 4.3). The difference in mass between the two sediments mirrors their difference in volume.

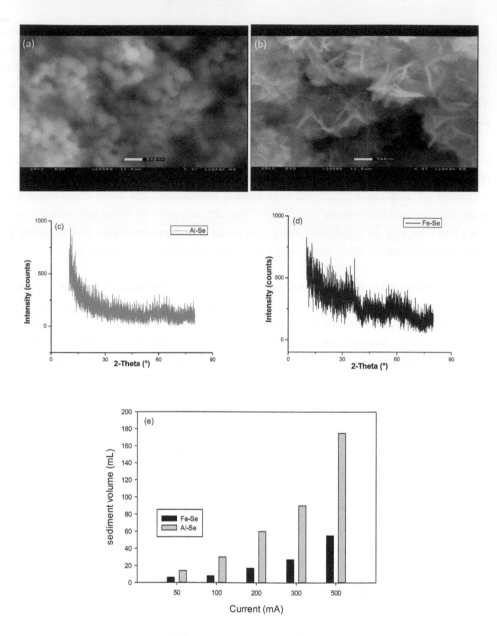

Figure 4.6. Sediment analysis: ESEM micrographs of (a) Se-Al sediment (500 nm scale), (b) Se-Fe sediment (500 nm scale), (c) XRD diffractogram of Se-Al sediment, (d) XRD diffractogram of Se-Fe sediment. Note: the sediments analyzed were produced at 100 mA under the following conditions [Conditions: Se(0) turbidity, 500 NTU; temperature, 20 °C; supporting media, 42 mmol L^{-1} NaCl; pH_0 = 7.0; 300 rpm; electrolysis time, 60 min], and (e) Evolution of Al-Se and Fe-Se sediment volume.

Table 4.3. Summary of Se-Fe and Se-Al sediment characteristics.

I (A)	Volume (mL)[a]	Volume (mL)[b]	Mass (g)[a]	Mass (g)[b]	Density (g/mL)[a]	Density (g/mL)[b]
0.05	5	11	9	16	1.74	1.45
0.1	12	28	20	38	1.64	1.36
0.2	24	58	31	75	1.28	1.29
0.3	31	82	36	106	1.16	1.29
0.5	38	170	42	188	1.11	1.11

[a]Fe-Se sediment, [b]Al-Se sediment

4.3.6. Electrical energy consumption and process optimization

The electrical energy consumption by the two kinds of electrodes is depicted on Figure 4.7. As can be seen in Table 4.3, for the first three applied currents, 50, 100 and 200 mA, Al and Fe electrodes had almost similar electrical energy consumption (below 1.68 kWh m^{-3}). Above 200 mA, at the highest turbidity removal efficiencies, Al electrodes showed a higher energy consumption than Fe (3.42 kWh m^{-3} versus 1.5 kWh m^{-3}).

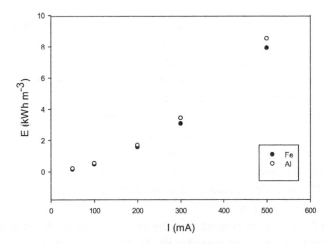

Figure 4.7. Electrical energy consumption in EC of Se(0) with Al or Fe electrodes.

Table 4.3 shows the importance of the process optimization. Using currents above the optimal value led to higher electrode consumption and more sediment generated. At the optimum Se(0) removal currents, Fe electrodes lost two times more of their mass comparing to Al electrodes: 0.417 kg m^{-3} versus 0.201 kg m^{-3}. On the other hand, the resulting Se-Al sediment (212 kg m^{-3}) was 3.4 times higher in weight than the Se-Fe sediment (61.6 kg m^{-3}). These contrasting results could be explained by the flocculation process (floc growth) which entraps water molecules within the metal-Se(0)-biopolymer matrix. Process optimization is thus a particularly important aspect when treating large volumes of Se(0) laden suspensions in a full-scale EC application.

Figure 4S. Treatment effectiveness of electrocoagulation. (a) Colloidal red Se(0) and (b) Electrocoagulation of colloidal Se(0) using Al and Fe electrodes. *Conditions*: 200 mA; pH$_0$ = 7.0; temperature, 20 °C; volume, 0.5 L; supporting electrolyte, 42 mmol L^{-1} of NaCl; 300 rpm; t = 60 min.

4.4. Conclusions

From the results obtained in this work the following conclusions can be drawn:
- Biogenic colloidal Se(0) can be effectively separated from aqueous solution by an electrocoagulation process using Fe or Al electrodes.
- Electrocoagulation with Al and Fe sacrificial electrodes (as anode) can achieve high (up to 97% with) removal efficiencies of colloidal Se(0).
- Low amounts of electrical energy (below 4 kWh m^{-3}) are consumed during EC of colloidal Se(0). Al electrodes consume twice the amount of energy required by the Fe electrodes to achieve comparable Se(0) removal efficiencies.
- At the highest colloidal Se(0) removal efficiency, the Fe electrodes are consumed two times faster than the Al electrodes. However, the resulted Se-Al sediment is 3.4 times more voluminous than the Se-Fe sediment.
- The TLCP of Se-Fe sediments suggest that the sediment can be listed as non-hazardous waste, whereas the Se-Al sediment exceeded the TCLP limit for Se by almost 20 times.

4.5. References

Akbal, F., & Camci, S. (2011). Copper, chromium and nickel removal from metal plating wastewater by electrocoagulation. *Desalination 269*(1-3), 214-222.

Al Aji, B., Yavuz, Y., & Koparal, A.S. (2012). Electrocoagulation of heavy metals containing model wastewater using monopolar iron electrodes. *Sep. Purif. Technol. 86*, 248-254.

Arredondo, M., & Nunez, M.T. (2005). Iron and copper metabolism. *Mol. Aspects Med. 26*, 313-327.

Buchs, B., Evangelou, M.W.H., Winkel, L.H.E., & Lenz, M. (2013). Colloidal properties of nanoparticular biogenic selenium govern environmental fate and bioremediation effectiveness. *Environ. Sci. Technol. 47*(5), 2401-2407.

Canizares, P., Carmona, M., Lobato, J., Martınez, F., & Rodrigo, M.A. (2005). Electrodissolution of aluminum electrodes in electrocoagulation processes. *Ind. Eng. Chem. Res. 44*, 4178-4185.

Canizares, P., Martınez, F., Jimenez, C., Lobato, J., & Rodrigo, M.A. (2007). Coagulation and electrocoagulation of wastes polluted with colloids. *Sep. Sci. Technol. 42*, 2157-2175.

Chapman, P.M., Adams, W.J., Brooks, M., Delos, C.G., Luoma, S.N., Maher, W.A., Ohlendorf, H.M., Presser, T.S., & Shaw, P. (2010). *Ecological assessment of selenium in the aquatic environments*. SETAC Press, Pensacola, Florida, USA.

Dobias, J., Suvorova. E.I., & Bernier-Latmani. R. (2011). Role of proteins in controlling selenium nanoparticle size. *Nanotechnology 22*(19), 195605.

Donald, A.M. (2003). The use of environmental scanning electron microscopy for imaging wet and insulating materials. *Nat. Mater. 2*, 511-513.

Dong, Y., Gandhi, N., & Hauschild, M.Z. (2014). Development of Comparative Toxicity Potentials of 14 cationic metals in freshwater. *Chemosphere 112*, 26-33.

Duan, J., & Gregory, J. (2003). Coagulation by hydrolyzing metal salts. *Adv. Colloid. Interfac. 100-102*, 475-502.

Gamage, N.P., & Chellam, S. (2011). Aluminum electrocoagulation pretreatment reduces fouling during surface water microfiltration. *J. Membrane Sci. 379*, 97-105.

Gensemer, R.W., & Playle, R.C. (1999). The bioavailability and toxicity of aluminum in aquatic environments. *Crit. Rev. Env. Sci. Technol. 29*(4), 315-450.

Hanai, O., & Hasar, H. (2011). Effect of anions on removing Cu^{2+}, Mn^{2+} and Zn^{2+} in electrocoagulation process using aluminum electrodes.*J. Hazard. Mater. 189*, 572-576.

Harif, T., & Adin, A. (2011). Size and structure evolution of kaolin-$Al(OH)_3$ flocs in the electroflocculation process: a study using static light scattering. *Water Res. 45*, 6195-6206.

Harif, T., Khai, M., & Adin, A. (2012). Electrocoagulation versus chemical coagulation: Coagulation/flocculation mechanisms and resulting floc characteristics. *Water Res. 4*, 3177-3188.

Heidmann, I., & Calmano, W. (2010). Removal of Ni, Cu and Cr from a galvanic wastewater in an electrocoagulation system with Fe- and Al-electrodes. *Sep. Purif. Technol. 71*, 308-314.

Holt PK, Barton GW, Wark M, & Mitchell CA (2002). A quantitative comparison between chemical dosing and electrocoagulation. *Colloid Surface A 211*, 233-248

Holt, P.K., Barton, G.W., & Mitchell, C.A. (2005). The future of electrocoagulation as a localized water treatment technology. *Chemosphere 59*, 335-367.

Jarvis, P., Jefferson, B., Gregory, J., & Parsons, S.A. (2005). A review of floc strength and breakage. *Water Res. 39*, 3121-3137.

Kabdasli, I., Arslan-Alaton, I., Olmez-Hanci, T., & Tunai, O. (2012). Electrocoagulation applications for industrial wastewaters: a critical review. *Environ. Technol. Rev. 1*, 2-45.

Khandegar, V., & Saroha, A.K. (2013). Electrocoagulation for the treatment of textile industry effluent - A review. *J. Environ. Manage. 128*, 949-963.

King, E.O., Ward, M.K., & Raney, D.E. (1954). Two simple media for the demonstration of pyocyanin and fluorescein. *J. Lab. Clin. Med. 44*, 301-307.

Lakshmanan, D., Clifford, D.A., & Samanta, G. (2009). Ferrous and ferric ion generation during iron electrocoagulation. *Environ. Sci. Technol. 43*, 3853-3859.

Lemly, A.D. (2002). Symptoms and implications of selenium toxicity in fish: the Belews Lake case example. *Aquat. Toxicol. 57*(1-2), 39-49.

Lemly, A.D. (2004). Aquatic selenium pollution is a global environmental safety issue. *Ecotox. Environ. Safe. 59*, 44-56.

Lenz, M., & Lens, P.N.L. (2009). The essential toxin: The changing perception of selenium in environmental sciences. *Sci. Total. Environ. 407*, 3620-3622.

Lenz, M., Kolvenbach, B., Gygax, B., Moes, S., & Corvinni, P.F.X. (2011). Shedding light on selenium biomineralization: proteins associated with bionanominerals. *Appl. Environ. Microb. 77*(13), 4676-4680.

Lide, D.R. (2004). *CRC Handbook of Chemistry and Physics,* 84th Edition. CRC Press. Boca Raton, Florida, US.

Luoma, S.N., Johns, C., Fisher, N.S., Steinberg, N.A., Oremland, R.S., & Reinfelder, J.R. (1992). Determination of selenium bioavailability to a bivalve from particulate and solute pathways. *Environ. Sci. Technol. 26*, 485-491.

Mechelhoff, M., Kelsall, G.H., & Graham, N.J.D. (2013). Super-faradaic charge yields for aluminum dissolution in neutral aqueous solutions. *Chem. Eng. Sci. 95*, 353-359.

Mello Ferreira, A., Marchesiello, M., & Thivel, P.X. (2013). Removal of copper, zinc and nickel present in natural water containing Ca^{2+} and HCO_3^- ions by electrocoagulation. *Sep. Purif. Technol. 107*, 109-117.

Mollah, M.Y.A., Schennach, R., Parga, J., & Cocke, D.L. (2001). Electrocoagulation (EC) – science and applications. *J. Hazard. Mater. B84*, 29-41.

Mollah, M.Y.A., Morkovsky, P., Gomes, J.A.G., Kesmez, M., Parga, J., & Cocke, D.L. (2004). Fundamental, present and future perspectives of electrocoagulation. *J. Hazard. Mater. 114*, 199-210.

Moreno, C.H.A., Cocke, D.L., Gomes, J.A.G., Morkovsky, P., Parga, J.R., Peterson, E., & Garcia, C. (2009). Electrochemical reactions for electrocoagulation using iron electrodes. *Ind. Eng. Chem. Res., 48*, 2275-2282.

Mouedhen, G., Feki, M., Wery, M.D.P., & Ayedi, H.F. (2008). Behavior of aluminum electrodes in electrocoagulation process. *J. Hazard. Mater. 150*, 124-135.

Ohlendorf, H.M. (1989). Bioaccumulation and effects of selenium in wildlife. In Jacobs, L.M. (Ed) *Selenium in agriculture and the environment. Am. S. Agron. S. Sci.,* Madison, Wisconsin, series number 23, pp. 133-177.

Öncel, M.S., Muhcu, A., Demirbas, E., & Kobya, M. (2013). A comparative study of chemical precipitation and electrocoagulation for treatment of coal acid drainage wastewater. *J. Environ. Chem. Eng. 1*, 989-995.

Picard, T., Cathalifaud-Feuillade, G., Mazet, M., & Vandensteendam, C. (2000). Cathodic dissolution in the electrocoagulation process using aluminium electrodes. *J. Environ. Monit. 2*, 77-80.

Presser, T.S., & Luoma, S.N. (2006). Forecasting selenium discharges to the San Francisco Bay-Delta Estuary: Ecological effects of a proposed San Luis Drain extension: U.S. Geological Survey Professional Paper 1646.

Purkerson, D.G., Doblin, M.A., Bollens, S.M., Luoma, S.N., & Cutter, G.A. (2003). Selenium in San Francisco Bay zooplankton: Potential effects of hydrodynamics and food web interactions. *Estuaries 26*, 956-969.

Richens, D.T. (1997). *The chemistry of aqua ions.* Wiley, Chichester, UK.

Roberge, P.R. (2008). *Corrosion Engineering: Principles and Practice.* McGraw-Hill Professional.

Schlekat, C.E., Dowdle, P.R., Lee, B.G., Luoma, S.N., & Oremland, R.S. (2000). Bioavailability of particle-associated selenium on the bivalve *Potamocorbila amuresis. Environ. Sci. Technol. 34*, 4504-4510.

Simmons, D.B., & Wallschlaeger, D. (2005). A critical review of the biogeochemistry and ecotoxicology of selenium in lotic and lentic environments. *Environ. Toxicol. Chem. 24*, 1331-1343.

Staicu, L.C., van Hullebusch, E.D., Oturan, M.A., Ackerson, C.J., & Lens, P.N.L. (2015a). Removal of colloidal biogenic selenium from wastewater. *Chemosphere 125*, 130-138.

Staicu, L.C., Ackerson, C.J., Cornelis, P., Ye, L., Berendsen, R.L., Hunter, W.J., Noblitt, S.D., Henry, C.S., Cappa, J.J., Montenieri, R.L., Wong, A.O., Musilova, L., Sura-de Jong, M., van Hullebusch, E.D., Lens, P.N.L., Pilon-Smits, E.A.H. (2015b). *Pseudomonas moraviensis* subsp. stanleyae: a bacterial endophyte capable of efficient selenite reduction to elemental selenium under aerobic conditions (*J. Appl. Microb.,* in revision).

Trompette, J.L., Vergnes, V., & Coufort, C. (2008). Enhanced electrocoagulation efficiency of lyophobic colloids in the presence of ammonium electrolytes. *Colloid. Surface. A 315*, 66-73.

Turchiuli, C., & Fargues, C. (2004). Influence of structural properties of alum and ferric flocs on sludge dewaterability. *Chem. Eng. J. 103*(1-3), 123-131.

USEPA.Test Methods for Evaluating Solid Waste, Physical/Chemical Methods, SW-846, Method 1311.Available from:

http://www.epa.gov/osw/hazard/testmethods/sw846/pdfs/1311.pdf

USEPA. Test Methods for Evaluating Solid Waste, Physical/Chemical Methods, SW-846, Method 1311, Chapter 7 – Characteristics, introduction and regulatory definitions. Available from:

http://www.epa.gov/wastes/hazard/testmethods/sw846/pdfs/chap7.pdf.

USEPA (1999). Enhanced coagulation and enhanced precipitative softening guidance manual. EPA 815-R-99-012.

USEPA (2010). North San Francisco Bay selenium characterization study plan (2010–2012). Available from: http://www2.epa.gov/sites/production/files/documents/epa-r09-ow-2010-0976-0023-1.pdf.

Un, U.T., Koparal, A.S., & Ogutveren, U.B. (2009). Hybrid processes for the treatment of cattle-slaughterhouse wastewater using Al and Fe electrodes. *J. Hazard. Mater. 164*, 580-586.

Vasudevan, S., Lakshmi, J., & Sozhan, G. (2009). Studies on the removal of iron from drinking water by electrocoagulation - A clean process. *Clean 37*, 45-51.

Vasudevan, S., & Oturan, M.A. (2014). Electrochemistry: as cause and cure in water pollution - an overview. *Environ. Chem. Lett. 12*, 97-108.

Zhang, Y., Zahir, Z.A., & Frankenberger Jr., W.T. (2004). Fate of colloidal-particulate elemental selenium in aquatic systems. *J. Environ. Qual. 33*, 559-564.

Zhu, B., Clifford, D.A., & Chellam, S. (2006). Comparison of electrocoagulation and chemical coagulation pretreatment for enhanced virus removal using microfiltration membranes. *Water Res.* 13, 3098-3108.

CHAPTER 5

Removal of colloidal biogenic selenium from wastewater

This chapter was published as:

Staicu, L.C., van Hullebusch, E.D., Oturan, M.A., Ackerson, C.J., & Lens, P.N.L. (2015). Removal of colloidal biogenic selenium from wastewater. *Chemosphere* 125, 130-138.

Chapter 5. Removal of colloidal biogenic selenium from wastewater

Abstract

Biogenic selenium, Se(0), has colloidal properties and thus poses solid-liquid separation problems, such as poor settling and membrane fouling. The separation of Se(0) from the bulk liquid was assessed by centrifugation, filtration, and coagulation-flocculation. Se(0) particles produced by an anaerobic granular sludge are normally distributed, ranging from 50 nm to 250 nm, with an average size of 166 ± 29 nm and a polydispersity index of 0.18. Due to its nanosize range and protein coating-associated negative zeta potential (-15 mV to -23 mV) between pH 2-12, biogenic Se(0) exhibits colloidal properties, hampering its removal from suspension. Centrifugation at different centrifugal speeds achieved 22 ± 3% (1,500 rpm), 73 ± 2% (3,000 rpm) and 91 ± 2% (4,500 rpm) removal. Separation by filtration through 0.45 µm filters resulted in 87 ± 1% Se(0) removal. Ferric chloride and aluminum sulfate were used as coagulants in coagulation-flocculation experiments. Aluminum sulfate achieved the highest turbidity removal (92 ± 2%) at a dose of 10^{-3} M, whereas ferric chloride achieved a maximum turbidity removal efficiency of only 43 ± 4% at 2.7×10^{-4} M. Charge repression plays a minor role in particle neutralization. The sediment volume resulting from $Al_2(SO_3)_4$ treatment is three times larger than that produced by $FeCl_3$.

Keywords: Elemental selenium; Colloids; Coagulation-flocculation; Aluminum sulfate; Filtration; Centrifugation.

5.1. Introduction

Selenium (Se) is a chalcogen element sharing common properties with sulfur (S) and tellurium (Te). Se has a complex biogeochemistry and is circulated through environmental compartments via both natural and anthropogenic processes (Chapman et al., 2010). Natural sources of selenium are crustal weathering and leaching, volcanism, sea salt spraying, and biological activities (Wen & Carignan, 2007). The anthropogenic release of selenium in the environment is mainly related to fossil fuel combustion, mining, non-metal smelting, and agriculture practiced on seleniferous soils (Lemly, 2004).

Of special interest is the very narrow window between selenium essentiality and toxicity (Levander & Burk, 2006). Based on blood plasma glutathione peroxidase activity as the selenium biomarker, a dietary reference intake (DRI) of 55 µg day^{-1} is proposed (IOM, 2000). In excess, selenium poisoning (i.e. selenosis) can result in hair loss, brittle nails and neurological pathologies (e.g. decreased cognitive function, convulsions, and weakness) (Tinggi, 2003). Estimated maximal intake levels of 910 µg Se day^{-1} caused by agriculture practiced on seleniferous soils were linked to endemic selenosis in China (Yang et al., 1989). The toxicity elicited by Se on biota is mainly related to the chemical speciation that selenium

106

undergoes under changing redox conditions. Amongst its oxidation states, Se oxyanions, namely selenite (Se[IV], SeO_3^{2-}) and selenate (Se[VI], SeO_4^{2-}), are water-soluble, bioavailable and toxic (Simmons & Wallschlaeger, 2005). In contrast, elemental selenium, Se(0), is solid and less toxic (Dungan & Frankenberger Jr., 1999). Nevertheless selenium nanoparticles (SeNPs) exhibited significant toxicity to mice (Zhang et al., 2005). In addition, particulate Se(0) has been reported to be bioavailable to bivalves (Luoma et al., 1992; Schlekat et al., 2000) and fish (Li et al., 2008). Furthermore, Se(0) is prone to re-oxidation to toxic SeO_3^{2-} and SeO_4^{2-} when discharged into aquatic ecosystems (Zhang et al., 2004).

A number of treatment technologies, including biological methods, aim to remove selenium oxyanions present in industrial wastewaters by reducing them to solid-phase elemental selenium (Lenz & Lens, 2009; Sobolewski, 2013). When biological treatment of selenium-laden wastewaters is performed, biogenic Se(0) is the solid end product that can be removed from the aqueous phase (Staicu et al., 2015). Due to its surface charge and nanometer size, biogenic Se(0) exhibits colloidal stability making its removal from the water phase difficult (Buchs et al., 2013). Coagulation-flocculation is one of the main processes employed both in drinking and wastewater treatment for the removal of colloidal and suspended particles (Duan & Gregory, 2003). The principle relies on the destabilization and settling of the colloids and suspended particles that cannot settle by gravity within practical time frames. Because of their proven efficiency and low cost, aluminum sulfate and ferric chloride are currently employed as coagulants on a large scale (Gregory & Duan, 2001). When coagulants are added to water, the metal ions (e.g. Al^{3+}, Fe^{3+}) hydrolyze spontaneously and form a series of metastable metal hydrolysis products (Richens, 1997). These metal hydrolysis products act upon the negatively charged particles held in suspension by hydrostatic repulsion forces (Russel et al., 1992). They alter the physical state of the suspended particles by repressing their charge (i.e. charge repression) and by forming large aggregates of $Al(OH)_3/Fe(OH)_3$ (i.e. sweep flocculation) which lead to particle sedimentation (Gregory and Duan, 2001). The use of filtration has been reported for removing colloidal particles other than Se(0). In a recent study, Johnson et al. (2014) has investigated the removal of particulate and colloidal silver in the sewage effluent discharged from several British wastewater treatment plants. On the other hand, centrifugation is rarely used for removing colloidal particles because it is an energy intensive process, but this approach can become feasible when treating highly turbid wastewaters (Thuvander et al., 2014).

Regardless of the utilization of coagulation-flocculation on a large scale for the removal of colloidal particles (Duan & Gregory, 2003), no systematic study has been done to investigate the separation of biologically-produced colloidal Se(0) from the bulk solution. The objectives of this study were, therefore, to characterize surface charge, stability and particle size distribution of biogenic Se(0) particles and to assess the solid-liquid separation potential of colloidal elemental selenium by filtration, centrifugation and coagulation-flocculation.

5.2. Materials and Methods

5.2.1. Chemicals and media

Sodium selenate, Na_2SeO_4, ≥98.0%, was purchased from Sigma Aldrich and fresh solutions were prepared before each experiment. All other reagents were of analytical grade. Ferric chloride hexahydrate, $FeCl_3.6H_2O$ (ACS reagent, > 98%) and aluminum sulfate octadecahydrate, $Al_2(SO_4)_3.18H_2O$ (ACS grade, > 98%) were purchased from Sigma Aldrich and Fischer Scientific, respectively. All solutions were prepared using deionized water.

Incubations were done using Basal Mineral Medium (BMM) containing (g L^{-1}): NH_4Cl (0.3), $NaCl$ (0.3), $CaCl_2.2H_2O$ (0.11), $MgCl_2.6H_2O$ (0.1), 1 mL L^{-1} acid trace element solution, 1 mL L^{-1} basic element solution, and 0.2 mg L^{-1} of vitamin solution (Stams et al., 1993). 10 mM sodium selenate and 20 mM lactate (as sodium L-lactate) were amended to the BMM.

5.2.2. Production of biogenic red Se(0)

BMM containing sodium selenate and 15 g L^{-1} (wet weight) inoculum was transferred to serum bottles. Anaerobic granular sludge sampled from an Upflow Anaerobic Sludge Blanket (UASB) reactor treating brewery wastewater was used as inoculum. The sludge was kindly provided by Biothane Systems International (Delft, the Netherlands) and the same sludge was used throughout all experiments. The inoculum had a Total Suspended Solids (TSS) and a Volatile Suspended Solids (VSS) content of, respectively, 54.6 g L^{-1} and 39.8 g L^{-1}, corresponding to a VSS/TSS ratio of 0.73 (Kijjanapanich et al., 2013).

The bottles were closed with butyl rubber septa and aluminum caps, the headspace was flushed with nitrogen gas for 15 min and the final headspace pressure adjusted to 1.7 bar (Astratinei et al., 2006). Incubation was performed at 30 °C, in the dark and under constant shaking at 100 rpm for 14 days.

At the end of the incubation period, the colorless Na_2SeO_4 solution had developed a red color (Figure 5.1a), indicative of biogenic Se(0) formation (Oremland et al., 2004). The Se(0) particles produced through microbial reduction of selenium oxyanions are designated 'biogenic' and represent the red allotrope of elemental selenium (Fellowes et al., 2011). After 14 days of incubation, bottles containing red Se(0) were left in a vertical position for 6 h allowing for the separation of the granular sludge inoculum from the bulk Se(0) solution. Se(0)-containing supernatant was carefully transferred to new recipients and used for coagulation experiments.

5.2.3. Se(0) protein-coating characterization

After sampling biogenic Se(0), the red suspension was centrifuged at 10,000 x g for 5 min and the pellet was washed three times in sterile Phosphate Buffer Saline (PBS). The proteins that were attached onto the biogenic Se(0) particles were denatured by the addition of 160 mM dithiothreitol (DTT) and 1% SDS (Sodium Dodecyl Sulfate) followed by boiling the samples at 95 °C in a water bath for 5 min. The denatured samples (60 µL) were loaded into 15% denaturing gels and run at constant current (30 mA) for 2 h in TBE (1x) buffer using Polyacrylamide Gel Electrophoresis (SDS-PAGE) (Laemmli, 1970). Coomassie™ Blue G-420 (limit of detection [LOD], 25 ng/band) was used to stain the protein bands.

5.2.4. Jar-test experiments

Coagulation tests were conducted in triplicate at room temperature (21 °C) using a Velp Scientifica 6 places flocculator (JLT6) in standard 500 mL glass beakers (88 mm internal diameter x 122 mm height) containing 300 mL of Se(0) suspension. The 25 mm high x 75 mm wide x 1 mm thick stirring paddles were immersed roughly at the middle height of the sample. The coagulation protocol followed the standard coagulation stages (adapted from USEPA, 1999): *rapid mixing* at 90 rpm (which corresponds to a mean velocity gradient, G = 55.7 s^{-1}) for 1 min, *slow mixing* (floc formation) at 20 rpm (G = 6.8 s^{-1}) for 25 min and *sedimentation* (no stirring) in graduated Imhoff cones for 60 min.

Ferric chloride and aluminum sulfate were employed for the chemical coagulation study. The coagulant concentration employed ranged from 0 (controls) to 10^{-3} M. This range was chosen based on previous studies on colloidal particle removal from humic substances, dissolved organic matter, kaolin model solutions and combined sewer overflow by coagulation-flocculation (Jung et al., 2005; Canizares et al., 2007; El Samrani et al., 2008).

Experiments were initiated by adding 10 mL of different concentrations of both coagulants (ferric chloride and aluminum sulfate) under agitation to every sample at a point below the free surface of the liquid (El Samrani et al., 2008). After 60 min of sedimentation, the sediment volume was read and 100 mL of supernatant was transferred to polyethylene sampling cups. Residual turbidity, conductivity, pH, and electrophoretic mobility measurements were performed shortly after. The sediment was recovered for further measurements, after the remaining supernatant was carefully siphoned with a glass pipette.

The following controls were included: (1) sterile treatments containing BMM and selenate without inoculum to assess the abiotic selenate conversion, (2) treatments containing BMM and inoculum to assess the turbidity produced by suspended microbial growth, (3) Se(0) treatments without coagulant to assess gravitational sedimentation of the Se(0) particles. Centrifugation of the biogenic Se(0) particles was performed using a Rotina 420 centrifuge with a 4784-A rotor accommodating 15 mL tubes at three centrifugal speeds: 1,500 rpm ($\approx$ 463 x g), 3,000 rpm ($\approx$ 1,851 x g), and 4,500 rpm ($\approx$ 4,166 x g). Filtration of the biogenic Se(0) particles was done using 0.45 µm (poly)tetrafluoroethylene (PTFE) syringe filters.

5.2.5. Analysis

Turbidity (Hach DR/890 Colorimeter) and pH (Metrohm 691 and a SenTix 21 WTW pH electrode) were measured at the beginning and at the end of the Jar-Test experiments. Turbidity wasexpressed in Nephelometric Turbidity Units (NTU) and measured according to USEPA (1999). The volume of settled Se(0) was measured in standard 1,000 mL Imhoff graduated cones (USEPA, 1999). Conductivity measurements were performed on a Radiometer Analytical MeterLab CDM 230.

Electrophoretic mobility measurements were performed on a Zetasizer Nano ZS (Malvern Instrument Ltd., Worchestershire, UK) using a laser beam at 633 nm and a scattering angle of 173° at 25 °C according to the manufacturer's instructions. The impact of pH on the Se(0) particle surface charge was investigated using an MPT-2 autotitrator upgraded to the Zetasizer Nano ZS.

For Transmission Electron Microscopy (TEM), samples were loaded into dialysis tubing (Fisher, 25 µm wall thickness, molecular weight cutoff = 3,500 Da) and dialyzed against 20 mM Tris solution (pH 7.4). The buffer was exchanged three times at three-hour intervals. 5 µL of a dialyzed sample were pipetted onto 400 mesh carbon coated copper TEM grids (EM Sciences). The resulting samples were examined in a JEOL JEM-1400 TEM operated at 100 kV and spot size 1. The Se(0) particle size distribution was determined by TEM image processing using ImageJ[TM] 1.47v software.

Capillary Suction Time (CST) was measured at room temperature with a Triton Model 200 CST Apparatus as described in the Instruction Manual and reported by Scholz (2005) using 5 mL of Se(0) suspension. Whatman no. 17 blotting paper was used in all CST measurements (Scholz, 2006).

5.2.6. Data analysis

All data were analyzed by using the data analysis software SigmaPlot 12.0v. The results are presented as average values and standard deviation (σ) of three independent experiments (triplicates, $n = 3$) unless otherwise stated. When the standard deviation values were smaller than 5%, the error bars were not represented. The Polydispersity Index (PDI) of Se(0) particles was calculated by dividing the standard deviation by the mean diameter of the particles.

5.3. Results

5.3.1. Biogenic Se(0) particle characterization

A representative TEM micrograph of red biogenic Se(0) solution (Figure 5.1a) is presented in Figure 5.1b. Biogenic elemental selenium consists of round-shaped and polydispersed particles (Figures 5.1b and 5.1c). The characteristics of Se(0) particles are compiled in Table 5.1, showing that the surface charge of the Se(0) particles exhibited a negative zeta potential value at neutral pH. Figure 5.1d presents the protein bands that were attached to the biogenic Se(0) particles. Based on the protein ladder used as a reference, the bands were characterized by molecular weights (in kDa) in the 115 to below 6 kDa range.

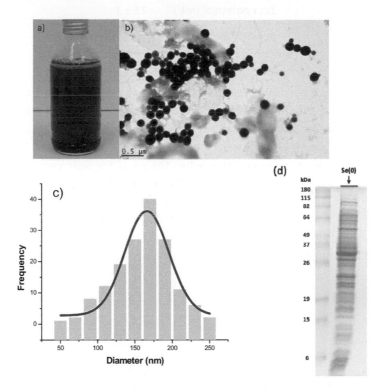

Figure 5.1. (a) Biogenic red Se(0) produced by anaerobic granular sludge inoculum (Incubation at: 30 °C, pH 7.5, anaerobic, dark, 100 rpm), (b) Transmission electron micrograph of biogenic red Se(0) produced by anaerobic granular sludge (same incubation conditions as (a)), (c) Size distribution of biogenic Se(0) particles. Note: n = 155, R^2 = 0.93, σ < 5%, and (d) SDS-PAGE gel image (Coomassie Blue G-420 stain) of proteins associated with biogenic Se(0) (right lane) and molecular standards in kilodaltons (left lane). Note that the arrow shows the red Se(0) that remained in the well after the gel was run.

Table 5.1. Characteristics of biogenic red Se(0) particles (incubated at 30 °C, initial pH, pH_0, = 7.5, anaerobic, dark, 100 rpm).

Parameter	Value
Shape	Spherical
Size range (nm)	50-250
Mean size (nm)	166 ± 29
Polydispersity index	0.18
Zeta potential (mV) [a]	- 23 ± 3
Point of Zero Charge	not reached

[a]measured at pH 7.7.

Figure 5.2 presents the biogenic Se(0) surface charge variation as a function of pH. In order to assess the influence of pH on the surface charge of Se(0) particles, the pH change was performed in both the acidic and basic domains starting from the initial pH of 7.7. Adding HCl or NaOH to a solution increases its ionic strength, which in turn acts by repressing the hydrodynamic diameter of the particles. When the pH of the Se(0) containing solution was increased, the initial zeta potential value (-23 mV, pH 7.7) also increased, reaching the highest value of -16.7 mV at pH 9.5. Above pH 9.5, the surface charge of Se(0) particles leveled off with a final zeta potential value of -17 mV at pH 12.5. Conversely, when the solution was acidified, the Se(0) particles displayed a higher increase in their surface charge reaching -13.8 mV at pH 2.6.

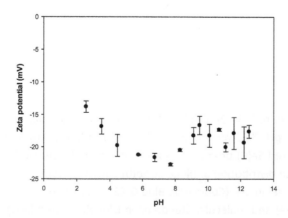

Figure 5.2. Variation of the zeta potential of biogenic Se(0) as a function of pH. Note that titration started at pH 7.7 and proceeded towards low and high pH.

Table 5.2 summarizes the characteristics of the Se(0) solution. The solution is characterized by high turbidity, deep red color (Figure 5.1a) and neutral pH. In contrast, the treatments without sodium selenate in the medium generated turbidity below 5 NTU after 14 days of incubation under similar conditions. Therefore, the high solution turbidity ($\approx$ 850 NTU) recorded for the incubations containing sodium selenate can be attributed to the red Se(0) observed both visually (Figure 5.1a) and with TEM (Figure 5.1b).

Table 5.2. Properties of biogenic Se(0) solution (produced at 30 °C, pH_0 = 7.5, anaerobic, dark, 100 rpm).

Parameter	Value
Turbidity (NTU)	850 ± 50
Color	Red
pH	7.5 ± 0.3
Conductivity (mS cm^{-1})	4.8 ± 0.2

Figure 5.3a shows the control treatment of Se(0) settling by gravitation over the time course of 4 days. The results are presented as normalized (C/C_0) colloidal stability where C represents the turbidity measured at different time points and C_0 is the initial turbidity. After a period of 4 days, biogenic Se(0) stored in Imhoff sedimentation cones exhibited only limited settling, 0.8 out of 1 (initial normalized turbidity), which corresponds to 179 NTU less turbidity.

5.3.2. Turbidity removal

Figure 5.3b depicts the Se(0) removal efficiency using centrifugation. The removal efficiency increased with centrifugal force: 1,500 rpm removed 22 ± 3% of colloidal Se(0), 3,000 rpm achieved 73 ± 2% and 4,500 rpm reached a Se(0) removal efficiency of 91 ± 2%. Filtration on a 0.45 µm cut-off filter removed 87 ± 1% of the initial Se(0).

The addition of aluminum sulfate improved the turbidity removal as a function of coagulant concentration added (Figure 5.4a). The lowest concentrations of coagulants (0.1 – 1.3 x 10^{-4} M) were not strong enough to impact the colloidal stability of biogenic Se(0) particles. When dosing 1.3 to 2.7 x 10^{-4} M aluminum sulfate, turbidity removal exhibited a marked increase from 2% to 57%, indicative of a threshold value that triggers the massive particle destabilization. Above 2.7 x 10^{-4} M, the decrease in turbidity was less pronounced until it reached about 92% removal efficiency for the highest coagulant dose used, 10^{-3} M.

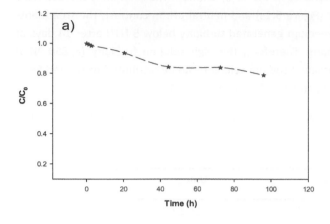

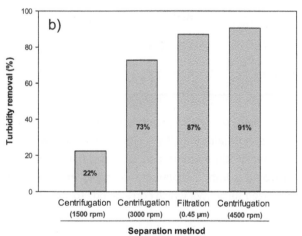

Figure 5.3. (a) Biogenic Se(0) gravitational settling (C – turbidity at time t, C_0 – initial turbidity) and (b) Removal of biogenic Se(0) using filtration and centrifugation at different speed values.

In contrast to aluminum sulfate, ferric chloride displayed a different turbidity removal curve and efficiency (Figure 5.4a). While it reached the threshold point (0.7×10^{-4} M) and the breakthrough domain of the curve faster than aluminum sulfate, the maximum removal efficiency achieved was only 43% for 4×10^{-4} M. Beyond this point, the removal efficiency undergoes a slow decline reaching 33% for 10^{-3} M. In terms of residual turbidity (Figure 5.4a and Table 5.2), the lowest value (70 NTU), was obtained for 10^{-3} M aluminum sulfate. Comparatively, the addition of ferric chloride could not decrease the turbidity below 458 NTU.

Figure 5.4b shows the change in pH over the coagulant dose range. Both treatments displayed decreasing pH with coagulant concentration, but the decrease was within one pH unit (from 8 to 7). Aluminum sulfate addition showed a more pronounced pH drop than ferric chloride.

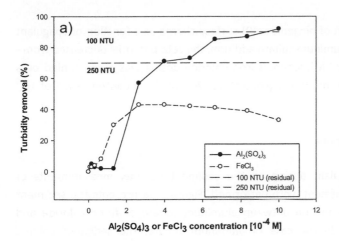

Figure 5.4. (a) Evolution of the Se(0) removal efficiency with coagulant dose. Note that 250 NTU and 100 NTU dashed lines indicate residual turbidity, (b) pH change of Se(0) solution as a function of coagulant dosage, and (c) Evolution of zeta potential with aluminum sulfate and ferric chloride concentration.

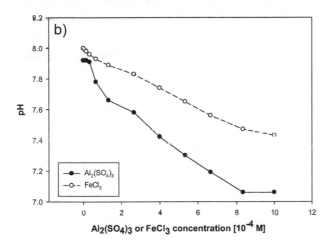

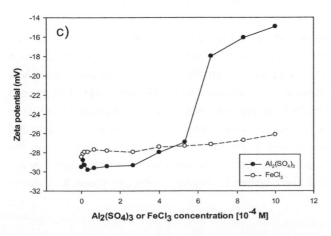

5.3.3. Se(0) surface charge

Figure 5.4c shows the evolution of biogenic Se(0) surface charge as a function of coagulant concentration. In the case of aluminum sulfate addition, the zeta potential decreased from -30 mV (control) to -15 mV (10^{-3} M), whereas iron chloride dosing was accompanied by a minor zeta potential change, from -28 mV (control) to -26 mV (for a concentration of 10^{-3} M).

5.3.4. Characterization of sediment

To clarify the effect of coagulant dosing on the sediment volume, measurements of sediment volume for each treatment were performed. Figure 5.5a presents the sediment volume resulting from the interaction between aluminum sulfate or ferric chloride and biogenic elemental selenium. In the case of aluminum sulfate, the sediment volume increased steadily reaching a maximum volume of 26 mL for the highest coagulant dose applied. In contrast, ferric chloride produced less sediment, showing a smooth increase until a maximum value of 8 mL for 10^{-3} M coagulant. For both coagulants, the first doses induced limited sedimentation. The sediment volume resulting from $Al_2(SO_3)_4$ treatment is about three times higher than that produced by $FeCl_3$.

Figure 5.5b presents the CST profile of the sediment collected during aluminum sulfate and ferric chloride treatment. For $Al_2(SO_4)_3$ treatment, the CST decreased with increasing coagulant dose, from 120 seconds to 87 seconds. The $FeCl_3$ treatment induced a reversed trend: the CST was increasing with coagulant dose from 86 seconds to 127 seconds.

5.4. Discussion

5.4.1. Turbidity removal

This study showed that aluminum sulfate is an effective coagulating agent to destabilize and sediment colloidal biogenic Se(0). In contrast, ferric chloride achieved only limited Se(0) removal. This difference might be related to the hydrolysis products of iron and aluminum obtained as a function of pH, which are significantly different. Even if both metals exhibit 4 deprotonations, from aquo Fe^{3+}/Al^{3+} complexesto $Fe(OH)_4^-$ / $Al(OH)_4^-$, aluminum hydrolysis species occur cooperatively and cover a much narrower pH range compared to iron species (Martin, 1991). This is explained by the transition from octahedral hexahydrate $Al^{3+}.6H_2O$ to tetrahedral $Al(OH)_4^-$, while in the case of iron its hydrolyzed species retain the octahedral co-ordination even if they span around 8 pH units (Martin, 1991). Apart from the mononuclear hydrolysis products, polymeric species can also play a role, although they are present in low concentrations and are inhibited by organic molecules (Duan & Gregory, 2003).

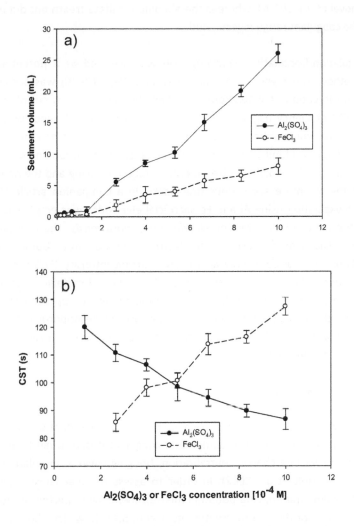

Figure 5.5. Sediment characterization (a) Effect of coagulant dose on sediment volume and (b) Evolution of CST of Al-Se and Fe-Se sediments as a function of coagulant concentration.

Flocculation is the second stage in chemical coagulation and has been shown to greatly impact the treatment effectiveness (Jarvis et al., 2005). Flocs are composed of a complex arrangement of solid particles, hydroxide precipitates and water entrapped during flocculation (Hogg, 2005). Floc growth is characterized by several phases until reaching a steady state between aggregation and fragmentation (Tambo, 1991). Ferric flocs were shown to be smaller and less compact than their aluminum sulfate counterparts, entailing also a poorer settleability (Turchiuli & Fargues, 2004). Ferric chloride displayed an Optimal Coagulant Concentration (OCC), defined as the coagulant dose that achieves the highest turbidity removal at 4×10^{-4} M, whereas the aluminum sulfate treatment did now show an OCC within the coagulant range investigated.

Prior to coagulation-flocculation, centrifugation was assessed as a potential solid-liquid separation method. The highest removal efficiency (91%) of Se(0) was achieved through high-speed centrifugation at 4,500 rpm. Filtration through 0.45 μm cut-off filters ranked second with 87% efficiency (Figure 5.3b). Although its removal efficiency was close to the performance of aluminum sulfate treatment, centrifugation cannot be employed on a full-scale situation due to prohibitive costs related to energy consumption and other capital and operating expenses. Additionally, pretreatment, membrane fouling and the need for further treatment of the retentate are serious drawbacks. In a companion article (Staicu et al., 2015), electrocoagulation using Al and Fe sacrificial electrodes was employed to sediment colloidal Se(0) produced by a strain of *Pseudomonas moraviensis*. The best Se(0) turbidity removal (97%) was achieved using iron electrodes at 200 mA. Aluminum electrodes removed 96% of colloidal Se(0) only at a higher current intensity (300 mA). While more efficient, electrocoagulation involves the use of electricity and metallic electrodes that are being consumed progressively. An interesting complementary approach to coagulation-flocculation or electrocoagulation might be the use of mesoporous silica conjugate adsorbents that have been shown to selectively detect and remove residual Se-oxyanions from aqueous media (Awual et al., 2014).

5.4.2. Se(0) charge repression

A number of coagulation mechanisms have been proposed dependent on the pollutant and physical-chemical properties of the solution and coagulant dosed (Duan & Gregory, 2003). These mechanisms include charge neutralization, bridging, sweep flocculation and double layer compression (Holt et al., 2002). In order to assess the mechanism involved in the biogenic Se(0) sedimentation, zeta potential measurements were performed for all applied coagulant doses and for the control treatments. Figure 5.4c shows that both coagulants act by repressing the surface charge (zeta potential) of Se(0) particles, but not up to the point of zero charge (PZC). This set of data indicates that charge repression contributes to the coagulation process to a lesser extent. Consequently, other coagulation mechanisms like sweep flocculation or adsorption appear to play the major role in the overall Se(0) removal by coagulant addition. However, the fact that both aluminum sulfate and ferric chloride acted by decreasing the surface charge of Se(0) particles should not be overlooked. Coagulation-flocculation is a complex phenomenon wherein several mechanisms are involved (Duan & Gregory, 2003). Similar results within the same pH range were reported for humic acid and kaolin suspensions (Zhao et al., 2011). Charge repression has been reported to act preferentially at acidic pH, whereas sweep flocculation and adsorption apply at neutral pH where abundant metal hydroxide precipitation is encountered (Edwards and Amirtharajan, 1985).

The change in pH (Figure 5.2) cannot repress the surface charge of Se(0) particles to the point of rendering them neutral. A possible explanation can be the protein layers that are coating biogenic Se(0) (nano)particles (Ni et al., 2014) and that act as a buffering system. It follows that the negatively charged protein/biopolymer layer coating of the Se(0) particles adds to the overall colloidal stability. Chemically-synthesized selenium particles have been shown to precipitate out from solution within minutes provided they are not stabilized with a capping agent (Johnson et al., 2008). In the case of chemogenic non-stabilized (nano)particles, pH is the key element that determines the surface charge and its magnitude (Uhm & Kim, 2014).

5.4.3. Sediment characterization

The volume of sediment formed is an important factor in a coagulation-flocculation approach. Sediment disposal and residuals management are serious challenges for wastewater treatment plants. Since both coagulants showed a delay in sedimentation for the lower doses employed (Figure 5.4a), this suggests the presence of a critical point in colloidal particle destabilization that should be reached in order to induce sedimentation. Beyond this point, the sedimentation depends on the nature and concentration of the coagulant but also on the colloidal particle characteristics (size, shape, surface charge, surface coating). Aluminum sulfate treatment produced three times more sediment than the $FeCl_3$ addition and this was consistent with the higher turbidity removal by the former. Interestingly, the decrease in turbidity removal beyond 2.7×10^{-4} M ferric chloride is not paralleled by a lower sediment volume. A possible explanation could be the composition of the sediments whereby $Fe(OH)_3$ and $Al(OH)_3$ have different contributions. Even if ferric chloride showed a lower efficiency in Se(0) coagulation, $Fe(OH)_3$ precipitates out from the solution and adds to the overall sediment volume.

Dewatering is a challenging task in coagulation-flocculation because less sediment volume means lower transportation and disposal costs. The Capillary Suction Time (CST) describes the tendency of the sediment to lose its water content and therefore reduce its volume (Scholz, 2005). The shorter the CST time (in seconds), the more prone the sediment is to dewatering. The opposite CST results (Figure 5.5b) observed in the current study can be linked to the floc structure of each sediment type. Fe-based sediments were shown to exhibit higher CST values and consequently a higher resistance to water removal than aluminum sulfate sediments (Turchiuli & Fargues, 2004). A possible explanation can be the smaller and less compact nature of iron flocs that allow them to retain water more tightly (Turchiuli & Fargues, 2004). Furthermore, the mineralogy and the structure of the sediments can be a key factor. Staicu et al. (2015) showed that Fe-Se sediments produced by electrocoagulation have a reticular structure, whereas Al-Se sediments lack an organized structure. These results can be linked to the lower dewaterability of Fe-Se sediments over their Al-Se counterpart.

The recovery and reuse of materials are important aspects of sustainable development. The resulted Al-Se and Fe-Se sediments could potentially be used as starting materials for Se recovery. Alternatively, since both Fe and Se are essential nutrients, the Fe-Se sediment might be considered as an amendment for Se deficient soils, e.g. in Ireland, India and Finland (Alfthan et al., 2014). Further investigation is required to fully assess the environmental impact of these sediments when applying Fe-Se sediments as potential fertilizers.

5.4.4. Biogenic Se(0) particle characteristics

Biogenic elemental selenium particles are produced by the metabolically driven reduction of selenium oxyanions (Oremland et al., 2004). Although a mass balance of the batch incubations was not made, the presence of residual Se oxyanions in the batch incubations is unlikely. A previous research using anaerobic granular sludge as inoculum and similar conditions as described in this paper (30 °C, pH = 7, anaerobic) gave an almost complete (97-99%) SeO_4^{2-} removal (Lenz et al., 2008). Moreover, possible intermediates of SeO_4^{2-} reduction (e.g. selenite; organic and inorganic selenides) were not detected in those incubations.

The difference in Se(0) size could be a result of the microbial community composition of the inoculum involved in the conversion of selenate. In addition, it has been shown that during stationary phase, *Shewanella* sp. HN-41 produces Se(0) particles with a normal size distribution (Tam et al., 2010). After 14 days of incubation, the inoculum generated Se(0) particles with a normal size distribution (Figure 5.1c). The polydispersity index ($\approx$ 0.18) suggests that even if the particle size ranges from 50 nm to 250 nm, the mechanism involved in biogenic Se(0) production has an important size control capacity. Considering a scale from 0 to 1, a PDI below 0.1 describes a high homogeneity of the particles, whereas high PDI values suggest a broader size distribution (Gaumet et al., 2008). The difference in Se(0) size could also be explained by the different cellular compartments where the particles are formed. The formation of Se(0) can take place in the cell envelope, reduction mediated by membrane-associated respiratory reductases, or inside the cell when toxic selenium oxyanions pass the outer respiratory barrier (Oremland et al., 2004). Moreover, since the Se(0) particles were produced anaerobically, elemental selenium granules can be stored and used when external electron acceptors become scarce (Herbel et al., 2003).

As a side investigation, the biogenic Se(0) particles were therefore subjected to denaturing gel electrophoresis in an attempt to confirm the presence of proteins attached to their surface (Figure 5.1d). The vast amount of protein bands that developed after staining may suggest the presence of a protein corona (Del Pino et al., 2014) and its potential involvement in the colloidal stability of the biogenic Se(0). Since Se(0) particle separation and purification by density gradient centrifugation were not performed, it is likely that some of the protein bands present in the gel image were not tightly associated with the Se(0) particles. Nevertheless, even present in large Se(0) and cell debris aggregates, the proteins add to the general colloidal stability. Similarly to other reports (Dobias et al., 2011; Lenz et al., 2011) focused on the identification of the proteins attached to biogenic Se(0) particles produced by pure cultures, it would be interesting to investigate the nature of the protein coating issued by mixed microbial cultures present in anaerobic granular sludges in a future study.

5.5. Conclusions

Biogenic Se(0) is colloidally stable between pH 2-12. The high colloidal stability is a direct consequence of the electrostatic forces associated with the Se(0)-protein coating that prevents aggregation and settling. Filtration, centrifugation and coagulation-flocculation can be employed for effective Se(0) particle sedimentation. Of these, coagulation-flocculation by aluminum sulfate has the best performance in the removal of Se(0)-associated turbidity. The difference in the Se(0) removal efficiency by coagulation-flocculation might be due to several factors, including the hydrolysis products of iron and aluminum, as well as floc size and structure. Charge repression of Se(0) particles showed only limited impact on the neutralization of the particle surface charge, suggesting that other mechanisms are likely to play a major role in Se(0) sedimentation. Due to possible differences in the mineralogical state of the two sediments, Al-Se and Fe-Se sediments display opposite dewatering characteristics. Al-Se sediments show a better dewatering potential that Fe-Se sediments. Highly turbid (800-900 NTU) wastewaters containing colloidal biogenic Se(0) can be effectively treated by coagulation-flocculation using aluminum sulfate.

5.6. References

Alfthan, G., Eurola, M., Ekholm, P., Venäläinen, E.R., Root, T., Korkalainen, K., Hartikainen, H., Salminen, P., Hietaniemi, V., Aspila, P., & Aro, A. (2014). Effects of nationwide addition of selenium to fertilizers on foods, and animal and human health in Finland: From deficiency to optimal selenium status of the population. *J. Trace Elem. Med. Biol.* doi: 10.1016/j.jtemb.2014.04.009.

Astratinei, V., van Hullebusch, E.D., & Lens, P.N.L. (2006). Bioconversion of selenate in methanogenic anaerobic granular sludge. *J. Environ. Qual.35*, 1873-1883.

Awual, M.R., Hasan, M.M., Ihara, T., & Yaita, T. (2014). Mesoporous silica based novel conjugate adsorbent for efficient selenium(IV) detection and removal from water. *Micropor. Mesopor. Mat. 197*, 331-338.

Buchs, B., Evangelou, M.W.H., Winkel, L.H.E., & Lenz, M. (2013). Colloidal properties of nanoparticular biogenic selenium govern environmental fate and bioremediation effectiveness. *Environ. Sci. Technol. 47*(5), 2401-2407.

Canizares, P., Martinez, F., Jimenez, C., Lobato, J., & Rodrigo, M.A. (2007). Coagulation and electrocoagulation of wastes polluted with colloids. *Separ. Sci. Technol. 42*, 2157-2175.

Chapman, P.M., Adams, W.J., Brooks, M., Delos, C.G., Luoma, S.N., Maher, W.A., Ohlendorf, H.M., Presser, T.S., & Shaw, P. (2010). *Ecological assessment of selenium in the aquatic environment*. SETAC Press, Pensacola, Florida, USA.

Del Pino, P., Pelaz, B., Zhang, Q., Maffre, P., Nienhaus, G.U., & Parak, W.J. (2014). Protein corona formation around nanoparticles - from the past to the future. *Mater. Horiz.1*, 301-313.

Dobias, J., Suvorova, E.I., & Bernier-Latmani, R. (2011). Role of proteins in controlling selenium nanoparticle size. *Nanotechnology 22*(19), 195605.

Duan, J., & Gregory, J. (2003). Coagulation by hydrolyzing metal salts. *Adv. Colloid. Interfac. 100-102*, 475-502.

Dungan, R.S., & Frankenberger Jr, W.T. (1999). Microbial transformations of selenium and the bioremediation of seleniferous environments. *Bioremed. J. 3*(3), 171-188.

Edwards, G.A., & Amirtharajah A. (1985). Removing color caused by humic acids. *J. Am. Water. Works.Ass. 77*, 50-57.

El Samrani, A.G., Lartiges, B.S., & Villieras, F. (2008).Chemical coagulation of combined sewer overflow: Heavy metal removal and treatment optimization. *Water Res. 42*, 951-960.

Fellowes, J.W., Pattrick, R.A.D., Green, D.I., Dent, A., Lloyd, J.R., & Pearce, C.I. (2011). Use of biogenic and abiotic elemental selenium nanospheres to sequester elemental mercury released from mercury contaminated museum specimens. *J. Hazard. Mater. 189*, 660-669.

Gaumet, M., Vargas, A., Gurny, R., & Delie, F. (2008). Nanoparticles for drug delivery: The need for precision in reporting particle size parameters. *Eur. J. Pharm. Biopharm. 69*(1), 1-9.

Gregory, J., & Duan, J. (2001). Hydrolyzing metal salts as coagulants. *Pure. Appl. Chem. 73*(12), 2017-2026.

Herbel, M.J., Switzer Blum, J., Borglin, S.E., & Oremland, R.S. (2003). Reduction of elemental selenium to selenide: experiments with anoxic sediments and bacteria that respire Se-oxyanions. *Geomicrobiol. J. 20*, 587-602.

Hogg, R. (2005). Flocculation and dewatering of fine-suspension particles. In: Stechemesser, H., & Dobias, B. (Ed.), *Coagulation and Flocculation*, 2nd ed., CRC Press, Florida, USA.

Holt, P.K., Barton, G.W., Wark, M., & Mitchell, C.A. (2002). A quantitative comparison between chemical dosing and electrocoagulation. *Colloid. Surface. A 211*, 233-248.

Houghton, J.I., Quarmby, J., & Stephenson, T. (2001). Municipal wastewater sludge dewaterability and the presence of microbial extracellular polymer. *Water Sci. Technol. 44*(2-3), 373-379.

Hundt, T.R., & O'Melia, C.R. (1988). Aluminum-fulvic acid interactions: mechanisms and applications. *J. Am. Water. Works. Ass. 80*(4), 176-186.

Institute of Medicine (IOM) (2000). *Dietary reference intakes for vitamin C, vitamin E, selenium, and carotenoids*. The National Academies Press, Washington, DC, USA.

Jarvis, P., Jefferson, B., Gregory, J., & Parsons, S.A. (2005). A review of floc strength and breakage. *Water Res. 39*(14), 3121-3137.

Johnson, N.C., Manchester, S., Sarin, L., Gao, Y., Kulaots, I., & Hurt, R.H. (2008). Mercury vapor release from broken compact fluorescent lamps and in situ capture by new nanomaterial sorbents. *Environ. Sci. Technol. 42*(15), 5772-5778.

Johnson, A.C., Jurgens, M.D., Lawlor, A.J., Cisowska, I., & Williams, R.J. (2014). Particulate and colloidal silver in sewage effluent and sludge discharged from British wastewater treatment plants. *Chemosphere 112*, 49-55.

Jung, A.V., Chanudet. V., Ghanbaja, J., Lartiges, B.S., & Bersillon., J.L. (2005). Coagulation of humic substances and dissolved organic matter with a ferric salt: An electron energy loss spectroscopy investigation. *Water Res. 39*(16), 3849-3862.

Kessi, J., Ramuz, M., Wehrli, E., Spycher, M., & Bachoefen, R. (1999). Reduction of selenite and detoxification of elemental selenium by the phototrophic bacterium *Rhodospirillum rubrum*. *Appl. Environ. Microbiol. 65*(11), 4734-4740.

Kijjanapanich, P., Annachhatre, A.P., Esposito, G., van Hullebusch, E.D., & Lens, P.N.L. (2013). Biological sulfate removal from gypsum contaminated construction and demolition debris. *J. Environ. Manage. 269*, 38-44.

Laemmli, U.K., (1970). Cleavage of structural proteins during the assembly of the head of bacteriophage T4. *Nature 227*, 680-685.

Lemly, A.D. (2004). Aquatic selenium pollution is a global environmental safety issue. *Ecotox. Environ. Safe. 59*, 44-56.

Lenz, M., & Lens, P.N.L. (2009). The essential toxin: The changing perception of selenium in environmental sciences. *Sci. Total. Environ. 407*, 3620-3633.

Lenz, M., Kolvenbach, B., Gygax, B., Moes, S., & Corvinni, P.F.X. (2011). Shedding light on selenium biomineralization: proteins associated with bionanominerals. *Appl. Environ. Microb. 77*(13), 4676-4680.

Levander, O.A., & Burk, R.F. (2006). Update of human dietary standards for selenium. In: Hatfield, D.L., Berry, M.J., & Gladyshev, V.N., (Ed.), *Selenium – its molecular biology and role in human health*. Springer, New York, USA.

Li, H., Zhang, J., Wang, T., Luo, W., Zhou, Q., & Jiang, G. (2008). Elemental selenium particles at nano-size (Nano-Se) are more toxic to Medaka (*Oryzias latipes*) as a consequence of hyper-accumulation of selenium: a comparison with sodium selenite. *Aquat. Toxicol. 89*(4), 251-256.

Luoma, S.N., Johns. C., Fisher. N.S., Steinberg. N.A., Oremland, R.S., & Reinfelder, J.R. (1992). Determination of selenium bioavailability to a bivalve from particulate and solute pathways. *Environ. Sci. Technol. 26*, 485-491.

Martin, R.B. (1991). Fe^{3+} and Al^{3+} hydrolysis equilibria. Cooperativity in Al^{3+} hydrolysis reactions. *J. Inorg. Biochem. 44*(2), 141-147.

Ni, T.W., Staicu, L.C., Schwartz, C., Crawford, D., Nemeth, R., Seligman, J., Hunter, W.J., Pilon-Smits, E.A.H., & Ackerson, C.J. (2015). Progress toward clonable inorganic nanoparticles (*Small*, in revision).

Oremland, R.S., Herbel, M.J., Blum, J.S., Langley, S., Beveridge, T.J., Ajayan, P.M., Sutto, T., Ellis, A.V., & Curran, S. (2004). Structural and spectral features of selenium nanospheres produced by Se-respiring bacteria. *Appl. Environ. Microb. 70*(1), 52-60.

Richens, D.T. (1997). *The chemistry of aqua ions*. Wiley, Chichester, UK.

Russel, W.B., Saville, D.A., & Schowalter, W.R. (1992). *Colloidal dispersions*. Cambridge University Press, UK.

Schlekat, C.E., Dowdle, P.R., Lee, B.G., Luoma, S.N., & Oremland, R.S. (2000). Bioavailability of particle-associated selenium on the bivalve *Potamocorbila amuresis*. *Environ. Sci. Technol. 34*, 4504-4510.

Scholz, M. (2005). Review of recent trends in Capillary Suction Time (CST) dewaterability testing research. *Ind. Eng. Chem. Res. 44*(22), 8157-8163.

Scholz, M. (2006). Revised capillary suction time (CST) test to reduce consumable costs and improve dewaterability interpretation. *J. Chem. Technol. Biot. 81*, 336-344.

Simmons, D.B., & Wallschlaeger, D. (2005). A critical review of the biogeochemistry and ecotoxicology of selenium in lotic and lentic environments. *Environ. Toxicol. Chem. 24*(6), 1331-1343.

Sobolewski, A. (2013). Evaluation of treatment options to reduce water-borne selenium at coal mines in West-Central Alberta. Microbial Technologies, Inc. Available from: http://environment.gov.ab.ca/info/library/7766.pdf.

Staicu, L.C., van Hullebusch, E.D., Lens, P.N.L., Pilon-Smits, E.A.H., & Oturan, M.A. (2015). Electrocoagulation of colloidal biogenic selenium. *Environ. Sci. Pollut. Res.Int. 22*(4), 3127-3137.

Stams, A.J.M., van Dijk, J.B., Dijkema, C., & Plugge, C.M. (1993). Growth of syntrophic propionate-oxidizing bacteria with fumarate in the absence of methanogenic bacteria. *Appl. Environ. Microb. 59*, 1114-1119.

Tam, K., Ho, C.T., Lee, J.H., Lai, M., Chang, C.H., Rheem, Y., Chen, W., Hur, H.G., & Myung, N.V. (2010). Growth mechanism of amorphous selenium nanoparticles synthesized by Shewanella sp. HN-41. *Biosci. Biotech. Biochem. 74*(4), 696-700.

Tinggi, U. (2003). Essentiality and toxicity of selenium and its status in Australia: a review. *Toxicol. Lett. 137*, 103–110.

Tambo, N. (1991). Basic concepts and innovative turn of coagulation/flocculation. *Water Supp. 9*, 1-10.

Thuvander, J., Arkell, A., & Jönsson, A.-S. (2014). Centrifugation as pretreatment before ultrafiltration of hemicelluloses extracted from wheat bran. *Sep. Purif. Technol. 138*, 1-6.

Turchiuli, C., & Fargues, C. (2004). Influence of structural properties of alum and ferric flocs on sludge dewaterability. *Chem. Eng. J. 103*(1-3), 123-131.

Uhm, H.N., & Kim, Y. (2014). Sensitivity of nanoparticles' stability at the point of zero charge (PZC). *J. Ind. Eng. Chem. 20* (5), 3175-3178.

USEPA (1999). Enhanced coagulation and enhanced precipitative softening guidance manual. EPA 815-R-99-012.

Wen, H., & Carignan, J. (2007). Reviews on atmospheric selenium: emissions, speciation and fate. *Atmos. Environ. 41*, 7151–65.

Yang, G., Yin, S., Zhou, R., Gu, L., Yan, B., Liu, Y., & Liu, Y.J. (1989). Studies of safe maximal daily dietary Se-intake in a seleniferous area in China. Part II: Relation between Se-intake and the manifestation of clinical signs and certain biochemical alterations in blood and urine. *Trace Elem. Electrolytes Health Dis. 3*, 123-130.

Zhang, Y., Zahir, Z.A., & Frankenberger Jr., W.T. (2004). Fate of colloidal-particulate elemental selenium in aquatic systems. *J. Environ. Qual. 33*, 559-564.

Zhang, E.S., Wang, H.L., Yan, X.X., & Zhang, L.D. (2005). Comparison of short-term toxicity between nano-Se and selenite in mice. *Life Sci. 76*, 1099-1109.

Zhao, Y.X., Gao, B.Y., Shon, H.K., Cao, B.C., & Kim, J.H. (2011). Coagulation characteristics of titanium (Ti) salt coagulant compared with aluminum (Al) and iron (Fe) salts. *J. Hazard. Mater. 185*, 1536-1542.

CHAPTER 6

CONCLUSIONS AND PERSPECTIVES

Production, recovery and reuse of biogenic elemental selenium

This chapter was published as:

Staicu, L.C., van Hullebusch, E.D., & Lens, P.N.L. (2015). Production, recovery and reuse of biogenic elemental selenium. *Environ. Chem. Lett.* 13(1), 89-96.

Chapter 6. Conclusions and Perspectives

Production, recovery and reuse of biogenic elemental selenium

Abstract

Selenium (Se) has caused several ecological disasters due to its toxicity and bioaccumulation along trophic networks. A variety of industrial activities that use or process fossil fuels and mineral ores (e.g. electricity generation, metal extraction and oil refining) generates wastewaters containing selenium. Currently, these wastewaters are insufficiently treated before being discharged in the environment. Several environmental biotechnological processes are employed to convert soluble selenium oxyanions, selenite and selenate, to solid elemental selenium, Se(0), as the latter state is considered less toxic. Applying a post-treatment solid-liquid separation step to these biological processes removes and separates Se(0) from the treated effluent. Here, we review the current state of the art of the sources of selenium rich waste streams and propose several approaches for the removal and reuse of this element. The major points are: (1) Biogenic Se(0) shows colloidal properties that can be offset by the addition of coagulants, either by dissolving multivalent salts or electrogenerating the coagulant *in situ*, (2) Recovered biogenic Se(0) is a secondary raw material, and (3) Biogenic Se(0) can be used for niche applications as fertilizers and adsorbent for metals. The biological treatment of industrial wastewater containing selenium can be linked with resource recovery as a sound and economic approach to alleviate the demand for this critical element.

Keywords: Selenium; Biogenic; Colloidal; Resource; Recovery; Reuse.

6.1. Environmental impact of selenium

Selenium (Se) is a chalcogen element with complex biogeochemistry (Fernandez-Martinez & Charlet, 2009). Se oxyanions, selenite (Se[IV], SeO_3^{2-}) and selenate (Se[VI], SeO_4^{2-}), have been documented to exhibit toxicity towards various aquatic groups (Simmons & Wallschlaeger, 2005). During the mid-1970s, chronic Se poisoning occurred as a result of Se oxyanions leached from ash deposited in the vicinity of Lake Belews (North Carolina) by a coal-fired power plant. The impact was devastating for the resident fish populations: 19 out of 20 species were eliminated (Lemly, 2002). Another major event occurred in the 1980s when agricultural drain water enriched in Se severely affected the migratory bird populations in Kesterson Reservoir, California (Ohlendorf, 1989).

Amongst its valence numbers, elemental selenium, Se(0), has long been considered as the least toxic due to its solid state and low bioavailability (Chapman et al., 2010). Se(0) has nevertheless been shown to be bioavailable to filter feeders and fish (Li et al., 2008; Luoma et al., 1992; Schlekat et al., 2000). Extensive research on the impact of colloidal Se(0) on filter-feeding mollusks and the trophic networks has been conducted in San Francisco Bay area (Schlekat et al., 2000; Purkerson et al., 2003; Presser & Luoma, 2006; USEPA, 2010). Furthermore, micro and nanosize Se(0) particles have increased reactivity caused by their high surface-volume ratio, leading to concerns about the short-term toxicity of Se(0) nanoparticles (SeNP) (Zhang et al., 2005). Moreover, Se(0) is reoxidized to SeO_3^{2-} and SeO_4^{2-} when Se(0) reaches an environment with an elevated redox potential, e.g. surface waters compared to anaerobic bioreactors or sediments (Zhang et al., 2004).

6.2. Biogenic Se(0) – Metabolism

Se(0) can be produced by chemical (i.e. chemogenic) or biological (i.e. biogenic) synthesis. Even if pure cultures have been shown to withstand high concentrations of toxic Se oxyanions (Hunter & Manter, 2011; Staicu et al., 2015b unpublished results), mixed microbial cultures, e.g. those present in anaerobic granular sludge, are used in full-scale bioreactors that treat large volumes of selenium-laden wastewaters (Pickett et al., 2008; Opara et al., 2014). Mixed cultures are preferred over pure cultures as wastewaters are not sterile, and thus have a contamination potential by microorganisms present in the wastewater. Besides, mixed cultures often grow in flocs, biofilm or granular sludges, which have an increased capacity to withstand toxicants (Schmidt & Ahring, 1996; van Hullebusch et al., 2003).

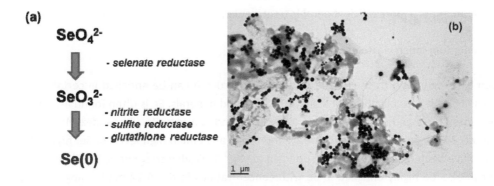

Figure 6.1. (a) Biological reduction of Se oxyanions to Se(0) and (b) Biogenic Se(0) produced by anaerobic granular sludge (Staicu et al., 2015c).

To date, only a limited number of bacteria (Lortie et al., 1992; Tomei et al., 1992; Tomei et al., 1994; Kuroda et al., 2011a) were shown to reduce SeO_4^{2-} to Se(0) (Fig. 6.1a), whereas the reduction of SeO_3^{2-} appears to be a more common feature of many phylogenetically diverse bacterial species (Sura-de Jong et al., 2014). Certain fungal species can also reduce selenite (Espinosa-Ortiz et al., 2015). The end product of Se oxyanions reduction can be Se(0) (Fig.

6.1b). Sometimes the complete reduction to Se^{2-} occurs, but this most reduced Se species is rapidly reoxidised to Se(0) under oxic conditions (Kagami et al., 2013).

The reduction of selenate is mediated by *SerA* reductase in *Thauera selenatis* (Schröder et al., 1997; Kraft et al., 2000) and by *SrdA* reductase in *Bacillus selenatarsenatis* (Kuroda et al., 2011b). A distinct selenate reductase, YnfE, was reported in *Citrobacter freundii* (Theisen & Yee, 2014). However, the exact biochemical mechanism of selenite reduction is still unclear. A vast amount of data suggests the presence of a common metabolic pathway used by bacteria for the reduction of other compounds (nitrate, nitrite, sulfate) as well (Fig. 6.1a). In the β-proteobacterium *Thauera selenatis*, a nitrite (NO_2^-) reductase has been reported to be involved in selenite reduction (DeMoll-Decker & Macy, 1993). In *Clostridium pasteurianum*, an inducible sulfite (SO_3^{2-}) reductase has been reported to reduce selenite pointing towards another putatively shared metabolic pathway (Harrison et al., 1980; Harrison et al., 1984).

Also, thiols have also been proposed to play a role in Se(0) formation. When amended with selenite, cultures of *Escherichia coli* (Bebien et al., 2002) and *Rhodospirillum rubrum* (Kessi & Hanselmann, 2004) showed overexpressed glutathione reductase activity, indicative of a detoxification process. Staicu et al. 2015b (unpublished results), reported a subspecies of *Pseudomonas moraviensis* able to withstand high levels of SeO_3^{2-} (up to 120 mM) and SeO_4^{2-} (over 150 mM). The strain was isolated from the roots of the Se hyperaccumulator plant *Stanleya pinnata* growing on seleniferous soils in Colorado (US). In addition, the strain can aerobically reduce 10 mM of SeO_3^{2-} below the detection limit within 48 h, with to red Se(0) as the end product. Because of its fast Se metabolism, the new isolate holds promise for the development of a more efficient aerobic treatment system of selenite-laden wastewaters.

6.3. Selenium-laden wastewater

Even if selenium is a trace element in the Earth's crust, it can be enriched in certain rocks and soils. As it can substitute for sulfur in biolites and minerals, Se is often found associated with fossil fuels and sulfide minerals, e.g. pyrite and chalcopyrite (Fernandez-Martinez & Charlet, 2009). Fossil fuels (e.g. crude oil and coal) can contain high levels of Se: high-sulfur coals have been reported to contain up to 43 mg Kg^{-1} Se (Yudovich & Ketris, 2006) and crude oil can also be highly enriched in Se with concentrations in the 5-22 mg L^{-1} range (Lemly, 2004).

Human activities like fossil fuel burning, non-metal smelting industry and agriculture practiced on Se-rich soils have resulted in the introduction of important quantities of selenium in the environment, thus modifying its natural cycle (Chapman et al., 2010). Wen and Carignan (2007) connected the increasing anthropogenic Se emissions to the onset of the Industrial Revolution (18th century). A positive correlation between the high coal combustion activity observed in 1940 and 1970 and the Se accumulation in archived herbage and soil samples from the Rothamsted Experimental Station (UK) has been documented (Haygarth et al., 1994). Considering the future trends in energy production based on fossil fuel combustion, it is expected that Se will increase its presence and toxicity in the environment (IEA, 2014).

By burning or processing Se-containing raw materials, wastewaters contaminated with high concentrations of this element are generated. Major sources of wastewaters containing selenium oxyanions are those of the oil refining industry, coal combustion and metal processing (Lemly, 2004; USEPA, 2010). Se laden wastewaters can be treated by a number of physical-chemical and biological treatment technologies (NAMC, 2010). Biological treatment processes using bioreactors produce different concentration levels of colloidal Se(0) as a function of the initial Se content and the Se conversion rate (NAMC, 2010). Several reactor types are currently employed on a full-scale configuration, notably Fluidized Bed Reactors (FBR) and the *Advanced Biological Metals Removal* (ABMet®) system. ABMet® employs anaerobic bacteria fixed to a granular activated carbon (GAC) bed or a biomatrix (Pickett et al., 2008).

6.4. Se(0) – Separation

Biogenic Se(0) has been shown to exhibit colloidal properties making it stable in suspension and difficult to sediment by gravitational settling (Buchs et al., 2013; Staicu et al., 2015b). The colloidal stability of biogenic Se(0) particles (Fig. 6.2) is related to the coating biopolymer layer that imparts an overall negative charge preventing the particles from aggregation (Buchs et al., 2013; Dobias et al., 2011). The chemogenic Se(0) undergoes allotropic transition shortly after synthesis unless not stabilized with a capping agent, e.g. protein and polyvinylpyrrolidone (Gates et al., 2002). This implies that the stability of chemogenic Se(0) is not intrinsic, but can be modified by adding stabilizing agents such as organic polymers (Jain et al., 2015b). Removal of colloidal Se(0) from a bioreactor effluent is necessary to reduce its environmental load and the potential negative impact of Se(0) exerted on aquatic ecosystems (See section 6.1). Several approaches have been investigated for the removal of colloidal Se(0), like chemical dosing (Staicu et al., 2015c) and electrocoagulation (Staicu et al., 2015b).

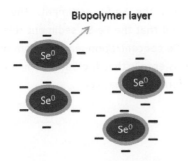

Biopolymer layer

Figure 6.2. Biopolymer coating that confers colloidal stability to biologically produced Se(0) nanoparticles.

6.4.1. Chemical dosing (Coagulation-Flocculation)

Coagulation-flocculation relies on the destabilization and settling of colloids and suspended particles that cannot sediment by gravity within practical time frames. When coagulants are added to water, the metal ions (e.g. Al^{3+}, Fe^{3+}) hydrolyze spontaneously and form a series of metastable metal hydrolysis products that alter the physical state of the suspended particles. The coagulants act by repressing the charge of the colloids and by forming large aggregates of $Al(OH)_3/Fe(OH)_3$ which ultimately leads to particle sedimentation (Gregory & Duan, 2001). Staicu et al. (2015b) investigated the effect of aluminum sulfate and ferric chloride on the solid-liquid separation of highly turbid (850 NTU) colloidal Se(0) solutions produced by anaerobic granular sludge under anaerobic conditions. Aluminum sulfate achieved the highest turbidity removal efficiency (92%) at a dose of 10^{-3} M, whereas ferric chloride achieved a maximum turbidity removal efficiency of only 43%. Charge repression appeared to play a minor role in particle neutralization.

6.4.2. Electrocoagulation

In electrocoagulation, an electrical current is applied between two electrodes (including a sacrificial anode) immersed in wastewater (Fig. 6.3a). Applying a current across the electrodes creates an electrical field and causes the dissolution of the sacrificial anode to form a coagulant and the electrolysis of water. The coagulants are thus electrogenerated *in situ* and in a continuous manner during the coagulation-sedimentation process. Another advantage of using an electrochemical approach for the treatment of wastewaters stems from the electrical field that exerts an electrophoretic action on the charged particles, inducing their migration towards the oppositely charged electrode.

Staicu et al. (2015a) reported the removal of colloidal Se(0) produced aerobically by *Pseudomonas moraviensis* stanleyae by electrocoagulation (Fig. 6.3b). Iron and aluminum sacrificial anodes were employed under galvanostatic (i.e. constant current) conditions. The best Se(0) turbidity removal efficiency (97%) was achieved using iron electrodes at 200 mA. Aluminum electrodes removed 96% of the colloidal Se(0) at a slightly higher current intensity (300 mA). Due to the less compact nature of the Al flocs (Fig. 6.3c), the Se-Al sediment was three times more voluminous than the Se-Fe sediment. The Toxicity Characteristic Leaching Procedure (TCLP) test showed that the Fe-Se sediment released Se below the regulatory level (1 mg L^{-1}), whereas the Se concentration leached from the Al-Se sediment exceeded the limit by about 20 times. This entails that the Se-Fe sediment might be landfilled, although recovery of Se is a more preferred alternative.

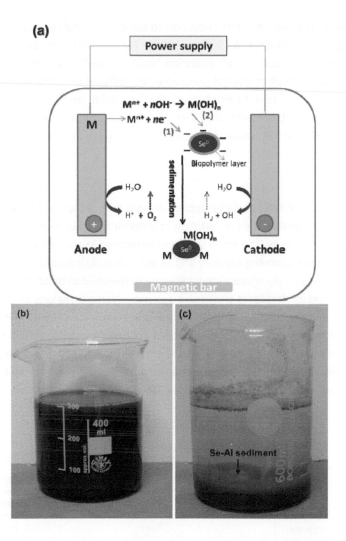

Figure 6.3. (a) Electrocoagulation setup and treatment effectiveness using Al sacrificial electrodes. Notes: M = metal (e.g. Al, Fe), (b) Colloidal biogenic Se(0) produced by *Pseudomonas moraviensis* stanleyae, and (c) Colloidal Se(0) treated by electrocoagulation with aluminum electrodes (Staicu et al., 2015a).

6.5. Recovery and reuse

Figure 6.4 presents a possible scheme of recovering and reusing Se oxyanions present in wastewaters. Biological treatment is a promising alternative that can reduce the load of Se that reaches the environment. In contrast to sulfur (S), Se can be converted from its oxidized forms directly to the Se(0) and thus Se removal can be achieved in a single reactor unit. The direct reduction of sulfur oxyanions to S(0) occurs in nature only in minor quantities. Therefore, high rate sulfur removal from wastewater needs to be achieved by a two-step process: first the complete reduction to sulfide (S^{2-}) and the subsequent oxidation to elemental sulfur (Muyzer & Stams, 2008). Furthermore, to avoid the complete reoxidation back to sulfate, the S^{2-} oxidation must proceed under strict microaerophilic conditions that are difficult to control (Janssen et al., 2001). If recycling is adopted, biogenic Se(0) particles, formed as a by-product of the biological treatment of selenium-laden wastewater, can link bioremediation with the generation of new materials. Several potential applications of the removed biogenic Se(0) are given below.

6.5.1. Fertilizers

The recovered biogenic Se(0) could be applied on soils that are low in bioavailable Se. Staicu et al. (2015a) have reported the precipitation of biogenic Se(0) with iron hydroxides using an electrocoagulation strategy. The Fe-Se sediment generated at the end of the treatment was shown to have a strong binding capacity and Se was slowly released from the iron matrix. A slower degradation and therefore a longer lifetime in the soil can be an argument in favor of using Se(0) instead of fertilizers containing selenate. Various regions of the world suffer from low Se (e.g. Finland, China, and Ireland) and in some countries (e.g. Finland) national programs have been successfully implemented to overcome this deficit (Alfthan et al., 2014). However, a detailed study is needed to investigate the behavior of Se(0) in agricultural soils. Other limitations, like social acceptance and the potential contamination of Se(0) with toxic metals co-present in wastewaters should also be addressed.

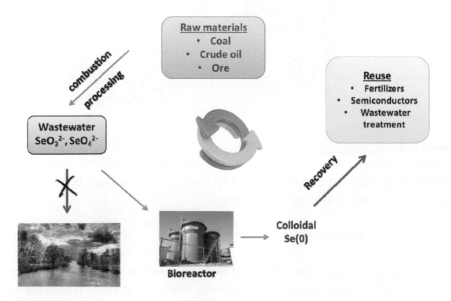

Figure 6.4. Proposed Se recovery and reuse scheme using Se rich wastewaters as a secondary Se resource.

6.5.2. Semiconductors

Having *photoconductive* properties (i.e. decreased electrical resistance with increased illumination) as well as *photovoltaic* properties (i.e. the direct conversion of light into electricity), metal-selenides are routinely exploited in a variety of industrial applications including solar and photo cells, exposure meters, and xerography (Johnson et al., 1999). The use of Se(0) as a starting material for the production of the II–VI semiconductors ZnSe and CdSe could be an alternative process for the production of metal-selenides. ZnSe nanoparticles were generated when selenite was reduced sequentially to Se(0) and then to selenide under anaerobic conditions by *Veillonella atypica* in the presence of $ZnCl_2$ (Pearce et al., 2008). The formation of metal-selenides through a disproportionation reaction between Se(0) and Ag^+ and Hg^{2+} has also been proposed (Nuttall, 1987).

6.5.3. Adsorbent for toxic metals

Another potential application for biogenic Se(0) is its use as adsorbent in the treatment of wastewaters and waste gases containing heavy metals (Jain et al., 2015a). Because biogenic Se(0) is covered by a biopolymer layer with an overall negative charge (Jain et al., 2015b), it will bind positively-charged ions from the bulk solution. Several studies were focused on the interaction of Se(0) with mercury (Hg^{2+}) (Johnson et al., 2008; Fellowes et al., 2011; Jiang et al., 2012) and zinc (Zn^{2+}) (Jain et al., 2015a). Se(0) was shown to form HgSe on the surface of Se(0) particles (Johnson et al., 2008). The retention of Hg in a solid phase is desirable from an ecotoxicological standpoint as the formation of toxic methyl mercury and volatile liquid Hg(0) is thus prevented. The Se(0) particles were shown to be around 20 times more efficient than commonly used adsorbents like nanosilver (188 vs 8.5 mg g^{-1}) and around 70 times than sulfur impregnated activated carbon nanotubes (188 vs 2.6 mg g^{-1}) (Johnson et al., 2008). It is interesting to note that the other nanoparticles tested (Cu, Ni, Zn), even if their size was in the nanorange domain, exhibited a limited adsorption capacity for Hg, pointing towards the influence induced by the biopolymer layer (Johnson et al., 2008).

The adsorption capacity of Zn^{2+} (added as $ZnCl_2$) onto biogenic Se(0) produced anaerobically by anaerobic granular sludge was recently quantified (Jain et al., 2015a). The maximum adsorption capacity amounted to 60 mg Zn g^{-1} biogenic Se(0). The effect of competing cations showed thatthe Zn^{2+} ion adsorption onto Se(0) was impacted by the presence of Ca^{2+} and Fe^{2+}, whereas the presence of Na^+ and Mg^{2+} had only a marginal impact. This is particularly relevant when treating industrial wastewaters, which usually have a complex matrix. As a comparison, zero valent iron (ZVI) has a much higher Zn adsorption capacity, 393 mg Zn/g ZVI (Yan et al., 2010). In order to gain more acceptance, the removal of soluble metals using Se(0) should be further investigated, particularly, with respect to the absorbance capacity of biogenic elemental Se and the stability of the metal-Se(0) matrices.

6.6. Perspectives

Considering the depletion of raw materials at an accelerated pace, the change of perception relative to the waste-resource dichotomy is an important pillar of sustainable development (Macaskie et al., 2010). Instead of discharging Se-laden wastewaters into the environment, Se can be converted into Se(0) through a bioremediation approach and recovered for further use. As there is limited incentive for such a strategy to become popular, legislation should be the major driving factor. Similarly to the removal of sulfur from flue gas streams under the pressure of the Clean Air Acts, selenium can become an element of great concern and increased interest. The biological treatment of Se-wastewaters is likely to gain in popularity in the coming years and so the amount of biogenic Se(0) produced and reused.

6.7. References

Alfthan, G., Eurola M., Ekholm, P., Venäläinen, E.R., Root, T., Korkalainen, K., Hartikainen, H., Salminen, P., Hietaniemi, V., Aspila, P., & Aro, A. (2014). Effects of nationwide addition of selenium to fertilizers on foods, and animal and human health in Finland: From deficiency to optimal selenium status of the population. *J. Trace Elem. Med. Biol.* doi: 10.1016/j.jtemb.2014.04.009.

Bebien, M., Lagniel, G., Garin, J., Touati, D., Vermeglio, A., & Labarre, J. (2002). Involvement of superoxide dismutases in the response of *Escherichia coli* to selenium oxides. *J. Bacteriol. 184*, 1556-1564.

Buchs, B., Evangelou, M.W.H., Winkel, L.H.E., & Lenz, M. (2013). Colloidal properties of nanoparticular biogenic selenium govern environmental fate and bioremediation effectiveness. *Environ. Sci. Technol. 47*(5), 2401-2407.

Chapman, P.M., Adams, W.J., Brooks, M., Delos, C.G., Luoma, S.N., Maher, W.A., Ohlendorf, H.M., Presser, T.S., & Shaw, P. (2010). *Ecological assessment of selenium in the aquatic environments*. SETAC Press, Pensacola, Florida, USA.

DeMoll-Decker, H., & Macy, J.M. (1993). The periplasmic nitrite reductase of *Thauera selenatis* may catalyze the reduction of selenite to elemental selenium. *Arch. Microbiol. 160*, 241-247.

Dobias, J., Suvorova. E.I., & Bernier-Latmani. R. (2011). Role of proteins in controlling selenium nanoparticle size. *Nanotechnology 22*(19), 195605.

Espinosa-Ortiz E.J., Gonzalez-Gil G., Saikaly P.E., van Hullebusch E.D. & Lens P.N.L. (2015). Effects of selenium oxyanions on the white-rot fungus *Phanerochaete chrysosporium. Appl. Environ. Microb. 99*(5), 2405-2418.

Fellowes, J.W., Pattrick, R.A., Green, D.I., Dent, A., Lloyd, J.R. & Pearce, C.I. (2011). Use of biogenic and abiotic elemental selenium nanospheres to sequester elemental mercury released from mercury contaminated museum specimens. *J. Hazard. Mater. 189*, 660-669.

Fernandez-Martinez, A. & Charlet, L. (2009). Selenium environmental cycling and bioavailability: a structural chemist point of view. *Rev. Environ. Sci. Biotechnol. 8*, 81-110.

Gates, B., Mayers, B., Cattle, B. & Xia, Y. (2002). Synthesis and characterization of uniform nanowires of trigonal selenium. *Adv. Funct. Mater. 12*, 219-227.

Gregory, J. & Duan, J. (2001). Hydrolyzing metal salts as coagulants. *Pure. Appl. Chem. 73*(12), 2017-2026.

Harrison, G.I., Laishley, E.J. & Krouse, H.R. (1980). Stable isotope fractionation by *Clostridium pasteurianum*. 3. Effect of SeO_3 on the physiology and associated sulfur isotope fractionation during SO_3 and SO_4 reductions. *Can. J. Microbiol. 26*, 952-958.

Harrison, G.I., Curle, C. & Laishley, E.J. (1984). Purification and characterization of an inducible dissimilatory type sulfite reductase from*Clostridium pasteurianum*. *Arch. Microbiol. 138*, 72–78.

Haygarth, P.M. (1994). Global importance and global cycling of selenium. In: Frankenberger, W.T. and Benson Sally, ed. *Selenium in the Environment*. New York, USA: Marcel Dekker; p. 1–28.

Hunter, W.J. & Manter, D.K. (2011).*Pseudomonas seleniipraecipitatus* sp. nov.: A selenite reducing γ-Proteobacteria isolated from soil. *Curr.MIcrobiol.62*, 565-569.

International Energy Agency (IEA) (2014). Available at: http://www.iea.org/topics/coal/

Jain, R., Jordan, N., Schild, D., van Hullebusch, E.D., Weiss, S., Franzen, C., Farges, F., & Lens, P.N.L. (2015a). Adsorption of zinc by biogenic elemental selenium nanoparticles. *Chem. Eng. J. 260*, 855-863.

Jain, R., Jordan, N., Weiss, S., Foerstendorf, H., Heim, K., Kacker, R., Hübner, R., Kramer, H., van Hullebusch, E.D., Farges, F. & Lens, P.N.L., (2015b). Extracellular polymeric substances (EPS) govern the surface charge of biogenic elemental selenium nanoparticles. *Environ. Sci. Technol. 49*(3), 1713-1720.

Janssen, A.J.H., Ruitenberg, R., & Buisman, C.J.(2001). Industrial applications of new sulphur biotechnology. *Water Sci. Technol. 44*, 85-90.

Jiang, S., Ho, C.T., Lee, J.H., Duong, H.V., Han, S., & Hur, H.G. (2012). Mercury capture into biogenic amorphous selenium nanospheres produced by mercury resistant *Shewanella putrefaciens* 200. *Chemosphere 87*, 621-624.

Johnson, J.A., Saboungi, M.L., Thiyagarajan, P., Csencsits, R., & Meisel, D. (1999). Selenium nanoparticles: a small-angle neutron scattering study. *J. Phys. Chem.B 103*, 59-63.

Johnson, N.C., Manchester, S., Sarin, L., Gao, Y., Kulaots, I., & Hurt, R.H.(2008). Mercury vapor release from broken compact fluorescent lamps and in situ capture by new nanomaterial sorbents. *Environ. Sci. Technol. 42*, 5772-5778.

Kagami, T., Narita, T., Kuroda, M., Notaguchi, E., Yamashita, M., Sei, K., Soda, S., & Ike, M. (2013). Effective selenium volatilization under aerobic conditions and recovery from the aqueous phase by Pseudomonas stutzeri NT-I. *Water Res. 47*, 1361-1368.

Kessi, J. & Hanselmann, K.W. (2004). Similarities between the abiotic reduction of selenite with glutathione and the dissimilatory reaction mediated by *Rhodospirillum rubrum* and *Escherichia coli*. *J. Biol. Chem. 279*(49), 50662-50669.

Krafft, T., Bowen, A., Theis, F., & Macy, J.M. (2000). Cloning and sequencing of the genes encoding the periplasmic-cytochrome B-containing selenate reductase of *Thauera selenatis*. *DNA Seq. 10* (6), 365-377.

Kuroda, M., Notaguchi, E., Sato, A., Yoshioka, M., Hasegawa, A., Kagami, T., Narita, T., Yamashita, M., Sei, K., Soda, S. & Ike, M. (2011a). Characterization of *Pseudomonas stutzeri* NT-I capable of removing soluble selenium from the aqueous phase under aerobic conditions. *J. Biosci. Bioeng.112*, 259-264.

Kuroda, M., Yamashita, M., Miwa, E., Imao, K., Fujimoto, N., Ono, H., Nagano, K., Sei, K., & Ike, M. (2011b). Molecular cloning and characterization of the srdBCA operon, encoding the respiratory selenate reductase complex, from the selenate-reducing bacterium *Bacillus selenatarsenatis* SF-1. *J. Bacteriol. 193*, 2141-2148.

Lemly, A.D. (2002). Symptoms and implications of selenium toxicity in fish: the Belews Lake case example. *Aquat. Toxicol. 57*(1-2), 39-49.

Lemly, A.D. (2004). Aquatic selenium pollution is a global environmental safety issue. *Ecotox. Environ. Safe.59*, 44-56.

Li, H., Zhang, J., Wang, T., Luo, W., Zhou, Q., & Jiang, G. (2008). Elemental selenium particles at nano-size (Nano-Se) are more toxic to Medaka (*Oryzias latipes*) as a consequence of hyper-accumulation of selenium: a comparison with sodium selenite. *Aquat. Toxicol. 89,* 251-256.

Lortie, L., Gould, W.D., Rajan, S., McCready, R.G.L. & Cheng, K.J. (1992). Reduction of selenate and selenite to elemental selenium by a *Pseudomonas stutzeri* isolate. *Appl. Environ. Microb. 58*, 4042-4044.

Luoma, S.N., Johns, C., Fisher, N.S., Steinberg, N.A., Oremland, R.S., & Reinfelder, J.R. (1992). Determination of selenium bioavailability to a bivalve from particulate and solute pathways.*Environ. Sci. Technol. 26*, 485-491.

Macaskie, L.E., Mikheenko,I.P., Yong, P., Deplanche, K., Murray, A.J., Paterson-Beedle, M., Coker, V.S., Pearce, C.I., Cutting, R., Pattrick, R.A.D., Vaughan, D., van der Laan, G., & Lloyd J.R. (2010). Today's wastes, tomorrow's materials for environmental protection. *Hydrometallurgy 104*, 483-487.

Muyzer, G. & Stams, A.J.M. (2008). The ecology and biotechnology of sulphate-reducing bacteria. *Nat. Rev. Microbiol. 6*, 441-454.

North American Metal Council (NAMC) (2010). Review of available technologies for removal of selenium from water. http://www.namc.org/docs/00062756.pdf

Nuttall, K.L. (1987). A model for metal selenide formation under biological conditions. *Med Hypotheses2*, 217-221.

Ohlendorf, H.M. (1989). Bioaccumulation and effects of selenium in wildlife. In Jacobs LM (ed) *Selenium in agriculture and the environment*. Am. S. Agron. S. Sci. Madison, Wisconsin, series number 23, pp. 133-177.

Opara, A., Peoples, M.J., Adams, J.D., & Martin, A.S. (2014). Electro-biochemical reactor (EBR) technology for selenium removal from British Columbia's coal-mining wastewaters.Available from: http://www.inotec.us/uploads/5/1/2/8/5128573/selenium_removal_coal_mine_water_inotec-sme2014.pdf

Pearce, C.I., Coker, V.S., Charnock, J.M., Pattrick, R.A., Mosselmans, J.F., Law, N., Beveridge, T.J., & Lloyd, J.R. (2008). Microbial manufacture of chalcogenide-based nanoparticles via the reduction of selenite using *Veillonella atypica*: an in situ EXAFS study. *Nanotechnology19*, 1-13.

Pickett, T., Sonstegard, J. & Bonkoski, B. (2008). Using biology to treat selenium. *Power Eng. 110*, 140-145.

Presser, T.S. & Luoma, S.N. (2006). Forecasting selenium discharges to the San Francisco Bay-Delta Estuary: Ecological effects of a proposed San Luis Drain extension. U.S. Geological Survey Professional Paper 1646.

Purkerson, D.G., Doblin, M.A., Bollens, S.M., Luoma, S.N., & Cutter, G.A. (2003). Selenium in San Francisco Bay zooplankton: Potential effects of hydrodynamics and food web interactions. *Estuaries 26*, 956-969.

Schmidt, J.E. & Ahring, B.K. (1996). Granular sludge formation in upflow anaerobic sludge blanket (UASB) reactors. *Biotechnol. Bioeng. 49*, 229-246.

Schroder, I., Rech, S., Krafft, T. & Macy, J.M. (1997). Purification and characterization of the selenate reductase from *Thauera selenatis. J. Biol. Chem. 272* (38), 23765-23768.

Simmons, D.B. & Wallschlaeger, D. (2005). A critical review of the biogeochemistry and ecotoxicology of selenium in lotic and lentic environments. *Environ. Toxicol. Chem.24*, 1331-1343.

Schlekat, C.E., Dowdle, P.R., Lee, B.G., Luoma, S.N., & Oremland, R.S. (2000). Bioavailability of particle-associated selenium on the bivalve *Potamocorbila amuresis. Environ. Sci. Technol. 34*, 4504-4510.

Staicu, L.C., van Hullebusch, E.D., Lens, P.N.L., Pilon-Smits, E.A.H., & Oturan, M.A. (2015a). Electrocoagulation of colloidal biogenic selenium. *Environ. Sci. Pollut. Res. Int.* 13(1), 89-96.

Staicu, L.C., Ackerson, C.J., Cornelis, P., Ye, L., Berendsen, R.L., Hunter, W.J., Noblitt, S.D., Henry, C.S., Cappa, J.J., Montenieri, R.L., Wong, A.O., Musilova, L., Sura-de Jong, M., van Hullebusch, E.D., Lens, P.N.L., & Pilon-Smits, E.A.H. (2015b, unpublished results). Pseudomonas moraviensis subsp. stanleyae: a bacterial endophyte capable of efficient selenite reduction to elemental selenium under aerobic conditions (*J. Appl. Microb.*, in revision).

Staicu, L.C., van Hullebusch, E.D., Oturan, M.A., Ackerson, C.J., & Lens, P.N.L. (2015c). Removal of colloidal biogenic selenium from wastewater. *Chemosphere 125*, 130-138.

Sura-de Jong, M., Reynolds,J., Richterova, K., Musilova, L., Staicu, L.C., Hrochova, I., Cappa, J.J., van der Lelie, D., Frantik, T., Sakmaryova, I., Strejcek, M., Cochran, A., Lovecka, P.

& Pilon-Smits, E.A.H. (2015). Selenium hyperaccumulators harbor a diverse endophytic bacterial community characterized by high selenium tolerance and growth promoting properties. *Front. Plant Sci. 6*, 113. doi: 10.3389/fpls.2015.00113.

Tomei, F.A., Barton, L.L., Lemanski, C.L. & Zocco, T.G. (1992). Reduction of selenate and selenite to elemental selenium by *Wolinella succinogenes*. *Can. J. Microb. 38*(12), 1328-1333.

Tomei, F.A., Barton, L.L., Lemanski, C.L., Zocco, T.G., Fink, N.H. & Sillerud, L.O. (1995). Transformation of selenate and selenite to elemental selenium by *Desulfovibrio desulfuricans*. *J. Ind. Microb. 14*(3-4), 329-336.

Theisen, J. & Yee, N. (2014). The molecular basis for selenate reduction in *Citrobacter freundii*. *Geomicrob. J. 31*, 875-883.

USEPA (2010). North San Francisco Bay selenium characterization study plan (2010–2012). http://www2.epa.gov/sites/production/files/documents/epa-r09-ow-2010-0976-0023-1.pdf

van Hullebusch, E.D., Zandvoort, M.H. & Lens, P.N.L. (2003). Metal immobilisation by biofilms: Mechanisms and analytical tools. *Rev. Environ. Sci. Biotechnol. 2*(1), 9-33.

Wen, H. & Carignan, J. (2007). Reviews on atmospheric selenium: emissions, speciation and fate. *Atmos. Environ. 41*, 7151-7165.

Yan, W., Herzing, A.A., Kiely, C.J., & Zhang, W.X. (2010). Nanoscale zero-valent iron (nZVI): Aspects of the core-shell structure and reactions with inorganic species in water. *J. Contam. Hydrol. 118*, 96-104.

Yudovich, Y.E. & Ketris, M.P. (2006). Selenium in coal: a review. *Int. J. Coal. Geol. 67*, 112-126.

Zhang, Y., Zahir, Z.A., & Frankenberger Jr., W.T. (2004). Fate of colloidal-particulate elemental selenium in aquatic systems. *J. Environ. Qual. 33*, 559-564.

Zhang, E.S., Wang, H.L., Yan, X.X., & Zhang, L.D. (2005). Comparison of short-term toxicity between Nano-Se and selenite in mice. *Life Sci. 76*, 1099-1109.

Appendix 1: Valorization of PhD research

I. Articles

1. Noblitt, S.D., **Staicu, L.C.**, Ackerson, C.J., & Henry, C.S. (2014). Sensitive, selective analysis of selenium oxoanions using microchip electrophoresis with contact conductivity detection. ***Anal. Chem.*** 86(16), 8425-8432.
2. **Staicu, L.C.**, van Hullebusch, E.D., Lens, P.N.L., Pilon-Smits, E.A.H., & Oturan, M.A. (2015). Electrocoagulation of colloidal biogenic selenium. ***Environ. Sci. Pollut. Res. Int.*** 22(4), 3127-3137.
3. **Staicu, L.C.**, van Hullebusch, E.D., Oturan, M.A., Ackerson, C.J., & Lens, P.N.L. (2015). Removal of colloidal biogenic selenium from wastewater. ***Chemosphere*** 125, 130-138.
4. **Staicu, L.C.**, van Hullebusch, E.D., & Lens, P.N.L. (2015). Production, recovery and reuse of biogenic elemental selenium. ***Environ. Chem. Lett.*** 13(1), 89-96.
5. Sura-de Jong, M., Reynolds, J., Richterova, K., Musilova, L., **Staicu, L.C.**, Hrochova, I., Cappa, J.J., van der Lelie, D., Frantik, T., Sakmaryova, I., Strejcek, M., Cochran, A., Lovecka, P. & Pilon-Smits, E.A.H. (2015). Selenium hyperaccumulators harbor a diverse endophytic bacterial community characterized by high selenium tolerance and growth promoting properties. ***Front. Plant Sci.*** 6,113.
6. **Staicu, L.C.**, Ackerson, C.J., Cornelis, P., Ye, L., Berendsen, R.L., Hunter, W.J., Noblitt, S.D., Henry, C.S., Cappa, J.J., Montenieri, R.L., Wong, A.O., Musilova, L., Sura-de Jong, M., Raynolds, J., van Hullebusch, E.D., Lens, P.N.L., & Pilon-Smits, E.A.H. (2015). *Pseudomonas moraviensis* subsp. stanleyae: a bacterial endophyte capable of efficient selenite reduction to elemental selenium under aerobic conditions. ***J. Appl. Microb.***, accepted.
7. **Staicu, L.C.***, Ni, T.W.*, Schwartz, C., Crawford, D., Nemeth, R., Seligman, J., Hunter, W.J., Pilon-Smits, E.A.H., & Ackerson, C.J. (2015). Progress toward clonable inorganic nanoparticles (***Small,*** in revision). ***Co-first authors**
8. **Staicu, L.C.**, van Hullebusch, E.D., & Lens, P.N.L. Treatment technologies for selenium-laden wastewaters (in preparation).

II. Conferences

Staicu, L.C., Ackerson, C.J., Cappa, J.J., & Pilon-Smits, EAH. (**2013**). Selenium biomineralization by *Variovorax paradoxus*. *4th International Symposium on Metallomics* (oral presentation). **Oviedo, Spain**.

Staicu, L.C., van Hullebusch, E.D., & Lens, P.N.L. (**2012**). Biorecovery of selenium from industrial wastewaters. *8th International Biometals Symposium* (oral presentation + poster). **Bruxelles, Belgium**.

Staicu, L.C., van Hullebusch, E.D., & Lens, P.N.L. (**2012**). Biorecovery of selenium from industrial wastewaters. *Microbial Sulfur Metabolism, E.M.B.O.* (oral presentation + poster). **Noordwijkerhout, The Netherlands**.

III. Summer school presentations

Staicu, L.C., van Hullebusch, E.D., Lens, P.N.L., & Oturan, M.A. (**2014**). Impact of electrocoagulation on biogenic colloidal selenium. *Paper presented at the ETeCoS³ summer school "Biological treatment of solid waste"*. **Cassino, Italy**. (http://www.internationaldoctorate.unicas.it/summerschool/etecos.html).

Staicu, L.C., van Hullebusch, E.D., & Lens, P.N.L. (**2012**). Biorecovery of selenium form industrial wastewaters. *Paper presented at the ETeCoS³ summer school "Contaminated soils: from characterization to remediation"*. University Paris Est, **Marne-la-Vallée, France**. (http://summer-schoolsoils.univ-paris-est.fr/).

Staicu, L.C., van Hullebusch, E.D., & Lens, P.N.L. (**2011**). Biorecovery of selenium from inorganic wastewaters.*Paper presented at the ETeCoS³ summer school "Biological and thermal treatment of municipal solid waste"*. **Naples, Italy.** (http://www.iat.unina.it/summerschool/home.html).

IV. Seminars

Staicu, L.C., van Hullebusch, E.D., Lens, P.N.L., & Oturan, M.A. (**2014**). Treatment of biogenic red selenium by electrocoagulation. *Paper presented at University Paris Est - PhD day*. **Marne-la-Vallée, France**.

Staicu, L.C., van Hullebusch, E.D., & Lens, P.N.L. (**2011**). Biorecovery of selenium from industrial wastewaters. *Paper presented at UNESCO-IHE PhD Week: Optimizing Water Usewith a Focus on Developing Countries*, **Delft**, **The Netherlands**.

V. Courses and trainings

"**Epistémologie et philosophie des sciences**", January-June **2014**, Paris Est University, given by Prof. Nicolas Bouleau, **Marne-la-Vallée, France**.

"**Environmental Metallomics**". July **2013**. Advanced course given by Prof. José L. Gómez Ariza (University of Huelva, Spain) and Prof. Andreas Prange (Helmholtz-Centre Geesthacht, Germany). **4th International Symposium on Metallomics**. **Oviedo, Spain**.

"**Fulbright enrichment seminar**". 13-16 December **2012**. **New Orleans, USA**.

"**Wastewater treatment technologies and modeling**". 11-15 June **2012**. Ph.D. course given by Prof. Ajit Annachhatre, Dr. Giovanni Esposito and Prof. Piet Lens. University Paris Est, **Marne-la-Vallée, France**.

"Microbial Community Engineering". February-April **2011**. Course given by Prof. Gerard Muyzer and Dr. Robbert Kleerebezem. Technical University of Delft (TU Delft), **Delft, The Netherlands**.

"Nanotechnology for water and wastewater". March 28 – April 8 **2011**. Course given by Prof. Piet Lens. UNESCO-IHE, **Delft, The Netherlands**.

Appendix 2: Curriculum vitae

Lucian STAICU

Born: September 22, 1981

E-mail: staiculucian@gmail.com
https://www.linkedin.com/in/lucianstaicu

EDUCATION

2010-14: **PhD in Environmental Technology**. UNESCO-IHE, Delft, The Netherlands.
2009-10: **Master Degree in Chemistry**. Paris Est Marne-la-Vallée University, France.
2008-10: **Master Degree in Environmental Research**. University of Bucharest, Romania.
2004-08: **Bachelor Degree in Biology and Ecology**. University of Bucharest, Romania.
2001-04: **Bachelor Degree in Physical Geography**. University of Bucharest, Romania.

RESEARCH ACTIVITIES

April 2015 – onwards: Postdoctoral researcher at Université de Franche-Comté, UFR Sciences et Techniques, Research group of Grégorio Crini. Besançon, France.

February 2015 - onwards: Researcher at Polytechnic University of Bucharest (UPB), Chemistry Department, Bucharest, Romania.

November 2010 - December 2014: PhD fellow position. **"Production of colloidal biogenic elemental selenium and removal by different coagulation-flocculation approaches"**. UNESCO-IHE (The Netherlands) and Paris Est University (France). Erasmus Mundus EteCoS[3] project. Promoter: Prof. Piet Lens (p.lens@unesco-ihe.org).

July 2013 - August 2014: **"Electrocoagulation of colloidal biofenis Se(0)"**. Uniersity Paris Est, France. Advisors: Prof. Mehmet Oturan (mehmet.oturan@univ-paris-est.fr) and Dr. Eric van Hullebusch (Eric.vanHullebusch@u-pem.fr).

August 2012 - June 2013: **"Microbial metabolism of selenium"**. Colorado State University, Biology Department, Fort Collins, Colorado, USA. Advisor: Prof. Elizabeth Pilon-Smits (epsmits@lamar.colostate.edu).

August 2012 - June 2013: "**Characterization of selenium nanoparticles**". Colorado State University, Chemistry Department, Fort Collins, Colorado, USA. Advisor: Prof. Christopher Ackerson (chris.ackerson@colostate.edu).

April 2013 - June 2013: "**Enzymes involved in the biomineralization of selenium**". United States Department of Agriculture (USDA), Fort Collins, Colorado, USA. Advisor: Dr. William Hunter (william.hunter@ars.usda.gov).

March 2010 - June 2010: M.Sc. research stage on Advanced Oxidation Processes. "**Degradation of mesotrione in aqueous medium using electro-Fenton process**", Paris Est University, France. Advisor: Prof. Mehmet Oturan (mehmet.oturan@univ-paris-est.fr).

FELLOWSHIPS AND GRANTS

Nov 2010 - August 2014. **Erasmus Mundus** EteCoS^3 PhD grant. UNESCO-IHE (**The Netherlands**) and University Paris Est (**France**).

August 2012 - June 2013. **Fulbright** Research Grant. Colorado State University, **USA**.

Oct 2009 - June 2010. **Agence Universitaire de la Francophonie** (AUF). Master 2. Paris Est University, **France**.

Sept 2008 - June 2010. MSc study grant awarded by University of Bucharest. Bucharest, **Romania**.

PERSONAL SKILLS AND COMPETENCES

Languages: **Romanian** (mother tongue), **English** (proficient), **French** (good user).

Language certificate: **TOEFL IBT**, November 5, 2011 (score: 109/120).

                            GRE Test Subject (**Biology**), April 4, 2009.

Technical skills and competences: **Microbiology** (bacterial isolation, biochemical and growth tests), **Genetics** (DNA electrophoresis, PCR), **Molecular Biology** (SDS-PAGE, Non-denaturing Gel Electrophoresis, Bradford, Western Blot, Enzyme assays), **Nanoparticles** (TEM, SEM, Zeta potential and size analysis, Chemical synthesis, Biogenic production), **Wastewater treatment** (Bioremediation, Electrochemical methods, Chemical dosing), **Electrochemistry** (Electrocoagulation, Advanced Oxidation Processes), **Analytical Chemistry** (Ion Chromatography, ICP, Spectrophotometry).

Computer skills and competences: Competent with most **Microsoft Office** programs (Word, Excel, PowerPoint), **Biostatistics** (OriginPro, SigmaPlot), **Scaffold4**, **LabVIEW**.

Research interests

Microbiology; Molecular Biology; Electrochemistry.

Netherlands Research School for the
Socio-Economic and Natural Sciences of the Environment

DIPLOMA

For specialised PhD training

The Netherlands Research School for the
Socio-Economic and Natural Sciences of the Environment
(SENSE) declares that

Lucian Constantin Staicu

born on 22 September 1981 in Bucharest, Romania

has successfully fulfilled all requirements of the
Educational Programme of SENSE.

Delft, 19 December 2014

the Chairman of the SENSE board

Prof. dr. Huub Rijnaarts

the SENSE Director of Education

Dr. Ad van Dommelen

The SENSE Research School has been accredited by the Royal Netherlands Academy of Arts and Sciences (KNAW)

KONINKLIJKE NEDERLANDSE
AKADEMIE VAN WETENSCHAPPEN

The SENSE Research School declares that Mr Lucian Staicu has successfully fulfilled all requirements of the Educational PhD Programme of SENSE with a work load of 51 EC, including the following activities:

<u>SENSE PhD Courses</u>

o Research Context Activity: Communicating research outcome in accessible press release: 'Scientists successfully precipitated colloidal selenium using electrocoagulation' (2014)

<u>Other PhD and Advanced MSc Courses</u>

o Nanotechnology for water and wastewater, UNESCO-IHE Delft (2011)
o Microbial Community Engineering, Technical University of Delft (2011)
o Ion Chromatography, UNESCO-IHE Delft (2011)
o Nanosizer, UNESCO-IHE Delft (2011)

<u>External training at a foreign research institute</u>

o Summer School 'Biological and Thermal Treatment of Municipal Solid Waste', Naples, Italy (2011)
o Summer school 'Contaminated Soils: From characterization to remediation' University Paris-Est, France (2012)
o Wastewater treatment technologies and modelling, University Paris-Est (2012)
o Training in 'Selenium metabolism of microbial isolates from various sources', New Orleans, United States (2012)
o Environmental Metallomics, Oviedo, Spain (2013)
o Molecular Biology and Nanotechnology, Colorado State University, United States (2013)
o Microtox, University Paris-Est, France (2014)
o Epistemology and philosophies of science, University Paris-Est (2014)
o Summer School 'Biological Treatment of Solid Waste', Cassino and Gaeta, Italy (2014)

<u>Oral Presentations</u>

o *Biorecovery of selenium from industrial wastewaters.* 8th International Biometals Symposium, 15-19 July 2012, Bruxelles, Belgium
o *Biomineralization of selenium in Variovorax paradoxus.* 4th International Symposium on Metallomics, 8-11 July 2013, Oviedo, Spain

SENSE Coordinator PhD Education

Dr. ing. Monique Gulickx

This work was supported by the European Union through the Erasmus Mundus Joint Doctorate Environmental Technologies for Contaminated Solids, Soils, and Sediments (ETeCoS³) programme (FPA n°2010-0009).

Cover picture: Red elemental selenium produced by *Pseudomonas moraviensis* stanleyae on KB medium.

Printed and bound by CPI Group (UK) Ltd, Croydon, CR0 4YY

21/10/2024

01777101-0014